Encountering Gorillas

C1

Encountering Gorillas

A Chronicle of Discovery, Exploitation, Understanding, and Survival

James L. Newman

ROWMAN & LITTLEFIELD PUBLISHERS, INC.
Lanham • Boulder • New York • Toronto • Plymouth, UK

Published by Rowman & Littlefield Publishers, Inc.
A wholly owned subsidiary of The Rowman & Littlefield Publishing Group, Inc.
4501 Forbes Boulevard, Suite 200, Lanham, Maryland 20706
www.rowman.com

10 Thornbury Road, Plymouth PL6 7PP, United Kingdom

British Library Cataloguing in Publication Information Available

Library of Congress Cataloging-in-Publication Data
Newman, James L.
Encountering gorillas : a chronicle of discovery, exploitation, understanding, and survival / James L. Newman.
pages cm
Includes bibliographical references and index.
ISBN 978-1-4422-1955-7 (cloth : alk. paper) — ISBN 978-1-4422-1957-1 (electronic)
1. Gorilla—Conservation. 2. Gorilla—Effect of human beings on. I. Title.
QL737.P96N49 2013
599.884—dc23

2013008124

™ The paper used in this publication meets the minimum requirements of American National Standard for Information Sciences—Permanence of Paper for Printed Library Materials, ANSI/NISO Z39.48-1992.

Printed in the United States of America

For Gorillas Everywhere

Contents

Illustrations

MAPS

PLATES

Preface and Acknowledgments

In search of an interesting elective during my undergraduate days at the University of Minnesota, I enrolled in a course titled "Exploration and Discovery." It didn't disappoint me, and I particularly enjoyed being able to travel back in time to visit distant lands and peoples. Over the ensuing years, I would occasionally read about explorers and their exploits, but serious study didn't begin until much later. In 1996, I had just completed a book called *The Peopling of Africa: A Geographic Interpretation*, dealing with events on the continent from prehistoric times up until the final decades of the nineteenth century. Explorers entered now and again, and I began to think about how they shaped the external world's view of Africa and impacted Europe's colonial undertakings. This led me to the famous—and, in many minds, infamous—Henry Morton Stanley and a 2004 book about him called *Imperial Footprints: Henry Morton Stanley's African Journeys*. While finishing it, a thought crossed my mind, namely, that true discovery isn't just a one-time thing but often a long-term process. By this I mean that once something is discovered, it is rediscovered as more information becomes available, and along with this discovery come changing views and impacts. Yes, I am aware that the idea of discovery is often ethnocentric. In Africa, people knew about mountains, lakes, rivers, gorillas, and so on before Europeans arrived to put them on a map. Exploration and discovery are thus for others, those living in the world beyond. As David Buisseret has remarked, it can be seen as "the process by which one or more people leave their society and venture to another part of the world (or, now, the heavens) then return in order to explain what they have seen."[1]

Somehow, and I can't remember why and when, gorillas entered my mind. I had been to Africa on various occasions and seen much of its iconic wildlife but not gorillas. Anyway, I began to read about them and in doing so realized they exemplified discovery as a continuing process. The outcome was a journal article called "The Discovery of Gorillas: The Journey from Myth to Reality."[2]

My thoughts, though, had become focused elsewhere. I had enjoyed researching and writing about Stanley and, after looking around for another subject, decided to try my hand at Richard Francis Burton, an undertaking that pretty much consumed my time for five years. With *Paths without Glory: Richard Francis Burton in Africa* finally behind me in 2010, I returned to thinking about gorillas. But how could I approach them? Clearly, it wouldn't be as either a professional primatologist or a conservationist. After some further reflection, I decided to expand on the idea of discovery by developing a story about how gorillas have entered the lives of people over time and how this has had consequences for them and, by extension, us as well. Furthermore, I had already written about people, and the project would enable me to do so again, this time dealing with an array of individuals who've been important in shaping how we think about and act toward gorillas. And these individuals would have to include gorillas themselves.

After making this decision, I had to decide on a title. *Gorillas in Our Midst* proved to be irresistible, as it played on Dian Fossey's *Gorillas in the Mist*, arguably the most famous book cum movie ever written about gorillas. But I soon discovered the title had been used previously for a small book about gorillas at the Columbus Zoo. Although titles aren't subject to copyright laws, I decided to search for another in order to avoid confusion. The current one eventually emerged as an even better option since the story is about encounters.

A subtitle clearly would be needed. "Discovery" had to be there, and as I began writing my narrative, it became crystal clear that gorillas had suffered terribly at the hands of humans, often as a result of misunderstandings due to bad press from many different quarters. It would take quite a while for the inaccuracies to be corrected, and as this happened, more and more people, specialists and nonspecialists alike, have rallied to their defense. Still, the survival of gorillas hangs in the balance.

For my books on Stanley and Burton, I relied heavily on archival documents, supplemented by an array of information from other sources, mostly in the physical and social sciences. I find the process of synthesizing diverse information stimulating. It's how my mind seems to work and thus how I've approached teaching and writing. This time my sources are mostly published works, along with movies and other representations, because public reception is crucial to the conditions gorillas find themselves in and also to attempts to ensure their continued survival. I've also ventured, with some trepidation, into the humanities and playing the role of movie critic.

I haven't done fieldwork among gorillas because I'm not equipped for this task. Specific training is required. I did consider going to Africa to visit them in either Rwanda or Uganda but decided not to. The cost is considerable, and looking at gorillas in one of the parks wouldn't yield anything positive beyond the thrill for me. I can't see the gorillas profiting by my presence. Moreover, I'm convinced that only qualified researchers and others engaged in protecting gorillas should be allowed near them. We pose many dangers, as I will try to show. Consequently, my visits to gorillas have been confined to zoos, where I also did some interviewing of staff for the section on zoos in the last chapter of the book.

I bring two perspectives to the study of gorillas different from previous works. As can be seen from my previous books noted above, I'm a student of Africa. It's where my mind has been focused throughout a career spanning more than forty years, with special attention paid to issues of population and environment. And gorillas are members of Africa's population who inhabit particular environments. Also, I'm a geographer by training, and while people of this persuasion study many different things from the inorganic to the organic, most agree that place matters. It can be a small place like a neighborhood, a region (such as the tropics), or even the whole world. It all depends on one's purposes. Along with this goes the notion that places are interconnected with others to varying degrees and that these interconnections change over time, altering the nature of places. In the following pages, we'll see how the places of gorillas have changed.

The story I've created is, for the most part, designed for general readers, especially but not limited to ones interested in the history of ideas, animal studies, wildlife conservation, exploration, and Africa. To this end, I've kept the details about gorilla biology and ecology limited to those necessary for an understanding of them and the conditions they face. In addition, I have not ventured into the areas of social theory, of which there is much regarding humans and animals. It's not my forte, and the language involved can be pretty dense and known only to other similarly inclined practitioners. Notes are provided for readers who wish to pursue these subjects. For gorilla specialists, I'm sure that parts of what I have to say will be familiar and, I trust, that I haven't done the details of their particular fields a disservice. On the other hand, if I've carried out my job in a proper fashion, specialists may find points of interest beyond their areas of expertise.

After chapter 1, "Introducing Gorillas," the story unfolds chronologically. It has to because the encounters have changed over time, with past ones influencing subsequent ones. Chapter 2 deals with early mythical encounters through the initial discovery of gorillas by the outside world in the mid-nineteenth century. A chapter follows on how their discovery set in motion a hunt for gorillas that took place in various ways and for various purposes. Discovery and its aftermath are continued in a chapter on Mountain gorillas, the best known to most people because of the attention they have received from adventurers, researchers, and the media. Chapter 5 is centered on how facts about gorillas began to replace old myths. Up to this point, people are at the center of attention, but in chapter 6, famous gorillas, both real and fictional, come to the fore. The book ends with a chapter on the prospects for gorilla survival both in their traditional homelands and in zoos.

Acknowledgments are always necessary since authors need others to help them find information and offer constructive comments. I have to start with Susan McEachern for her enthusiastic support of my proposal and invaluable editorial assistance. Others at Rowman & Littlefield I owe are Janice Braunstein, Carolyn Broadwell-Tkach, and Bruce R. Owens. Then there's Joe Stoll, cartographer and image maker supreme. As always, Marc Leo Felix can be counted on for sage advice, especially about African art. I owe many thanks to Jeroen Stevens for arranging my visit to the Antwerp Zoo and then for sharing so much information with me. And while there, I had the good fortune to speak with Cameroonian researcher Denis Ndeloh Etiendem about Cross River gorillas. I want to single out Beth Behner and her colleagues for their efforts to make sure I had up-to-date information about the Philadelphia Zoo. Baseball addict and on-the-scene-reporter Russ Tarby deserves a nod for encouragement and for comments on my movie reviews. And Ashley Swynnerton went out of her way to arrange a private screening at the Museum of Modern Art in New York City of the rare 1926 film *Gorilla Hunter* by Ben Burbridge. I owe debts of gratitude to Marilyn Silberfein for finding contacts at the Philadelphia Zoo and Mathilde Leduc Grimaldi for tracking down images of *Gorille enlevant une Femme*. Many thanks go to Dennis McCourt for his translation of German.

Many others helped smooth the way for me. They include Mai Qaraman Reitmeyer at the American Museum of Natural History; the Barcelona Zoo for sending images of the gorilla Snowflake almost instantly; Grahm Jones and Audra Meinelt at the Columbus Zoo; Ashley Morton at the National Geographic Society; Dana Lombardo, Kim Lengel, and Lynn Tunmer at the Philadelphia Zoo; anonymous persons at the Woodland Park Zoo; and, not to be forgotten, Photofest for its amazing poster images. And once again, I can't forget the staff of Interlibrary Loan at the Byrd Library of Syracuse University for locating some very hard to find publications. Last, but hardly least, there's Carole, who has put up with my distractions and long disappearances into my study cum messy cave.

NOTES

1. David Buisseret, ed., *The Oxford Companion to World Exploration* (New York: Oxford University Press, 2007), xxiii.
2. James L. Newman, "Discovering Gorillas: The Journey from Mythic to Real," *Terrae Incognitae* 38 (2006): 36–54. Permission to republish portions of the article kindly granted by *Terrae Incognitae* and Maney Publishing.

Chapter One

Introducing Gorillas

Gorillas have been portrayed in many different ways. Explorers, adventurers, fiction writers, artists, filmmakers, and scientists have all taken the opportunity to voice their views. In the pages that follow, I'll be drawing on much of what they've said and illustrated in order to create a different kind of story. It's about the ways gorillas and we humans have been brought together over time and how this has impacted their lives and, reciprocally, ours. Four themes, those of discovery, exploitation, understanding, and survival (?), undergird what I have to say. The question mark after survival has a twofold meaning: will gorillas survive, and, if so, where and how? Although it is true that much life on earth can be looked at through similar kinds of lenses, gorillas, in many ways, constitute a special case. They are close kin biologically; they're charismatic, meaning possessed of a power to command attention; until relatively recent times, they occupied highly specialized habitats where few, if any, people lived; and, as noted, they face a highly uncertain future. Making the story even more compelling is the fact that their precarious situation in today's world is owed to us. Some of this is due to often-told myths and misrepresentations, matching or even superseding those about wolves, snakes, and sharks. Only slowly did research begin to reveal the truth about gorillas, although a few of the distortions from the past linger on. Today, it's more a matter of gorillas being directly confronted by humans, causing them to lose territory due to displacement and induced environmental changes.

How one reacts to the plight of gorillas depends to a great extent on how gorillas are perceived. We can see a larger version of ourselves or see a quite different kind of being. Some may view gorillas as ugly and fearsome creatures and thus to be avoided, whereas others are drawn to them. They might be considered pests or thought of as one of nature's wonders. Who the person is makes a difference. How primatologists and conservationists view gorillas is unlikely to mirror the perspective of an African farmer whose fields are raided by them or be the same as the CEO of a firm harvesting tropical hardwoods. Each has different priorities, which, in turn, influence their thoughts and actions and, consequently, the lives of gorillas.

Even within Africa, the vast majority of people have never seen a live gorilla, much less one in the wild. Consequently, whatever comes to mind is most likely the product of words and pictures provided by others. In this regard, four rather remarkable and quite different people—Paul Du Chaillu, Carl Akeley, George Schaller, and Dian Fossey—head the list of people who brought gorillas to public and scientific attention. The supporting cast, though, is substantial and includes individual gorillas in addition to humans. The most famous live gorilla was Gargantua, seen by many thousands of people who attended the shows of the Ringling Brothers and Barnum & Bailey Combined Shows from the late 1930s until after World War II. And we must not forget fictional gorillas, headed by King Kong, who, in three incarnations, has thrilled and in the process informed and misinformed audiences for nearly eighty years. Artistic representations and museum dioramas have also played important roles in shaping what people see, know, and think about gorillas. Given the large number of other contributors, I'll focus attention on those who have been the most prominent or interesting in bringing humans and gorillas closer together.

Some basics about gorillas need to be established before my story begins.[1] They are by far the largest of the so-called anthropoids, or great apes (family Hominidae), along with orangutans (*Pongo pygmaeus* and *Pongo abelii*), common chimpanzees (*Pan troglodytes*), and bonobos (*Pan paniscus*). The four varieties of gibbons were once included with them, but they have been reclassified as belonging to the family Hylobatidae and often referred to as lesser apes. Many early descriptions portrayed gorillas as not just large but huge. One male, for example, was credited with being seven and a half feet tall and topping the scales at 770 pounds.[2] In fact, free-living adult males are slightly over five feet tall, with the heaviest weighing around 450 pounds. Only in zoos have they reached 600 pounds or more. In general, full-grown females weigh about half that of their male counterparts.

Origins are always difficult to pinpoint, but a good place in time to begin the temporal journey of becoming gorillas is within closed evergreen forests as the Oligocene Epoch gave way to the Miocene around 26 million years ago. Products of prior global warming, forests covered much of the land surface from southern Europe and Africa through India to China and Southeast Asia, and up in the trees resided primates galore. With full-color, stereoscopic vision and an ability to grasp things with fingers and thumbs, primates found a wide array of arboreal niches at their disposal and thrived as never before.

Many of the species that came into being had relatively short life histories and left no known descendants. Some, however, were members of the line leading to Old World monkeys, whereas still others served as ancestors to apes.

The first ape, assuming that only one such type existed, has yet to be identified, and there is also no agreement on how many Oligocene and Miocene apes there were—maybe two dozen or so, depending on how they are grouped.[3] *Proconsul*, *Aegyptopithecus*, *Pliopithecus*, *Dryopithecus*, *Graecopithecus*, *Oranopithecus*, *Sivapithecus*, and *Ramapithecus* are among the better-described representatives, but even their evolutionary positions remain undetermined because scientists have little to study other than teeth and jaw fragments. At best, we can probably say that early apes were fairly generalized types that had not yet become adapted for either swinging through trees (brachiation) or knuckle walking on the ground, the distinctive modes of locomotion for current apes.

Although several authorities have argued for an Asian origin of apes, the evidence supports the view that they were confined to Africa until around 15 million to 17 million years ago, when some migrated across land bridges into Eurasia, perhaps pushed by the onset of drier and cooler times that reduced the extent of their original forest homes.[4] In Southeast Asia they eventually produced gibbons and orangutans, both essentially arboreal, whereas in Africa evolution led to apes coming down from but not necessarily out of the trees. Two distinct primate realms emerged—monkeys inhabited the upper reaches of the forests, whereas apes exploited food sources closer to the ground. Whichever line of African ape might have been the precursor, about 9 million years ago gorillas separated from it, claiming the understories of forests, with a preference for edge habitats. Several million years later, chimpanzees, bonobos, and humans began their distinctive evolutionary journeys. Bonobos, like gorillas, remained within forested areas, whereas chimpanzees and the ancestors of humans ventured into woodlands and more open country.

How to classify gorillas is debatable. As is usual, some taxonomists are lumpers, others splitters.[5] Today a single genus, *Gorilla*, and four subspecies are generally recognized, but differences exist over the matter of whether there is one species or two. Favoring the latter are western and eastern groupings that are separated from one another by nearly 700 miles, and recent preliminary DNA analyses suggest that the genetic distance between them is greater than is found between the common chimpanzee and the bonobo. Consequently, those advocating two species have the floor, at least for now.

The most widespread subspecies is *Gorilla gorilla gorilla*, found in forest pockets from the Sanaga River in Cameroon on into the Republic of the Congo (Brazzaville) westward to the Oubangui River. A small population has recently been proven to exist within the Angolan enclave of Cabinda.[6] The

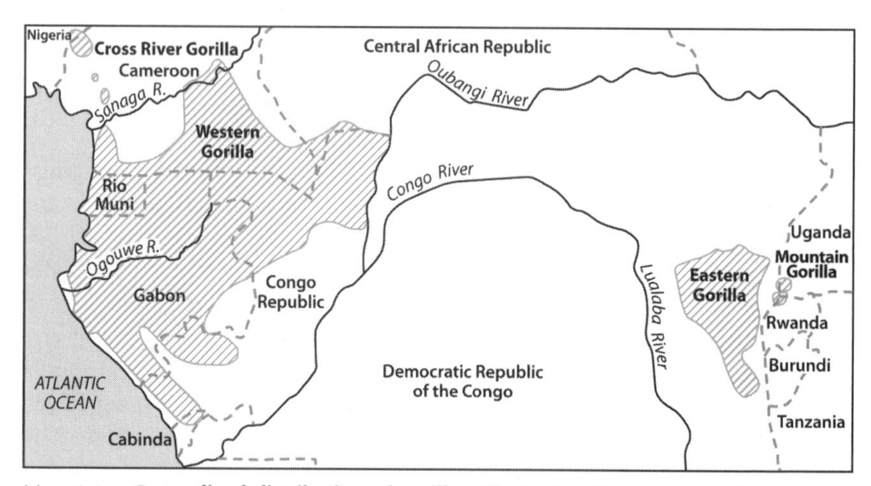

Map 1.1. Generalized distribution of gorillas, illustrating the gap separating western and eastern populations.

total number of *Gorilla gorilla gorilla* may be as high as 100,000; however reliable data are rare due to the large area over which they range and the difficulty of finding groups within dense tropical forest settings. Counting sleeping nests is the most widely used census method, and thus any figure you see must be considered a rough estimate at best.

Those unfamiliar with taxonomy may wonder why there are three instances of "gorilla" in the name. The reason is that they were the first of the four subspecies to be identified, and subsequent ones needed their own species or subspecies identities. Thus, in the semi-deciduous and evergreen forests within the rugged uplands bordering the Cross River in southeastern Nigeria and adjacent Cameroon live, at best estimate, about 300 to 350 individuals of *Gorilla gorilla diehli.* They were so named by mammologist Paul Matchie in 1903 after a certain Mr. Diehl, an employee of the German Northwestern Cameroon Company.[7] In subsequent years, they faded from view until a letter from a Dr. N. A. Dyce Sharp appeared in the 1929 edition of *Nature*, followed two years later by G. J. Allen's short article in an obscure journal, *The Nigerian Field*.[8] As happened previously, the gorillas then pretty much disappeared from official view until being "rediscovered" in the late 1980s. Some authorities, in fact, previously thought they had become extinct. By this time, too, taxonomists generally classified them as *Gorilla gorilla gorilla*. This has changed in response to recent physical measurements showing that Cross River gorillas have smaller teeth, palates, and cranial vaults, plus shorter skulls, than their southern neighbors.[9] Matchie has, thus, been vindicated. Based on this evidence, it seems likely the two populations became separated by a gap in the forest created during a Pleistocene dry phase many thousands of years ago. Where the few gorillas located in the Ebo Forest on the north bank of the Sanaga River fit has yet to be determined, although preliminary evidence points to the likelihood of their being an outlier of *Gorilla gorilla gorilla.*[10]

Within the Virunga Mountains, a string of eight volcanoes towering above the Albert rift valley slicing through Rwanda and adjacent portions of the Democratic Republic of the Congo (DRC) and Uganda, plus the nearby Bwindi Impenetrable Forest, live nearly 900 of the much more rare (although better known to most people) Mountain variety, *Gorilla beringei beringei.* The Virunga and Bwindi populations have been separated from one another for a very long time, and further research is needed to see if it's appropriate to place them within the same subspecies.

West of the Mountain gorillas in the DRC reside groups of *Gorilla beringei graueri*, first identified by Rudolf Grauer in 1908 and given separate status by Matchie in 1914. As will be seen later, several expeditions in the 1920s and 1930s obtained specimens, thinking them to be Mountain gorillas. In the early 1990s, about 17,000 individuals were thought to exist, but shortly afterward, the Rwandan crisis and civil war in the DRC resulted in an untold number being killed when refugees flooded into their forest homes. Consequently, no one knows just how many still survive. It could be in the neighborhood of several thousand or considerably fewer.

The often-used "eastern lowland gorilla" designation is not really apt because some troops live at higher elevations than their mountain brethren. Because no other common name has come into usage, I will refer to them as Eastern gorillas. In order to maintain a degree of geographic consistency, the others will henceforth be designated Western, Cross River, and Mountain.

How did the wide gap between gorillas in the west and those in the east come to be? The most popular explanation has modern gorillas originating in the west and spreading eastward through a continuous belt of lowland forest that once spanned much of the continent north of the equator. When cooler and drier conditions developed around 3 million years ago, woodlands and savannas intruded into the forest, eventually cutting off east-to-west contact. Such open country is hostile to gorillas since adults consume enormous quantities of fruits, leaves, roots, bulbs, shoots, flowers, and bark on a daily basis that only dense forests can provide. These forests also provide the security that troops seek. Later, in the eastern region, some groups migrated into the forests of the Virunga volcanoes and adjacent highlands, eventually reaching as high as 13,000 feet in elevation, whereas others remained at lower elevations.

Yet movement could also have taken place in the opposite direction, with gorillas coming down from the mountains and migrating through the lowland forests toward the Atlantic coast. Evidence to support this theory includes morphological characteristics, such as large barrel chests and relatively short limbs that are suggestive of an original adaptation to cold, high-altitude conditions, and the much greater gorilla population densities in montane than in lowland forests.[11] Given the current state of knowledge, which lacks fossil evidence, there's just not enough to go on to tell which interpretation comes closest to being the truth.

Based on an array of circumstantial evidence, it's been posited that the gap separating the western and eastern populations may have been smaller in the not-too-distant past. A tale from the Basoko of the Ituri region of the DRC, where no gorillas currently exist, is interesting in this regard. Called "A Hospitable Gorilla," it tells of a fisherman who one day came across a large male gorilla, which, after examining the man's hands, determined they were kin and gave him a password to use so that he would be recognized as such by other gorillas.[12] A little while later, the fisherman joined a hunting party, and, as chance would have it, he encountered the large gorilla with his family rushing toward him in an effort to escape. So as to avoid being trampled, the fisherman shouted the password loud and clear, and upon hearing it, the gorilla called out "Ah, this is our friend. Do not hurt him." A young male bringing up the rear failed to heed the command, and as he was about to attack the fisherman, the large male returned and slew the offender. Later, in another encounter, the gorilla provided the fisherman with a feast from the forest, saying a "wise man need not starve." The fisherman went back to tell the villagers the good news, and as a result they all agreed "that in the future the gorillas should be reckoned among those, against whom it would not be lawful to raise their spears." Should there have been gorillas there at the time, some humans obviously later raised spears.

Then there are the so-called Bondo gorillas from the vicinity of the Uele River valley. These first came to light with the discovery of four skulls that have the sagittal crest (a bone ridge running from front to back along the crest of the skullcap to which powerful jaw muscles are attached) characteristic of male gorillas and in other ways show a resemblance to the Western version.[13] Since then, several expeditions have documented the presence of some kind of unique ape population residing here, more likely of chimpanzee than of gorilla lineage.[14]

In 1898, Captain Guy Burrows published a photograph with the caption "Gorilla Killed at Stanley Falls." Close inspection reveals a large male chimpanzee, although at the time some authorities accepted it as a gorilla.[15] While exploring the Ituri region in 1905, George Grenfell, a Baptist missionary, reported seeing a gorilla family sitting in trees. The African men with him wanted fresh meat, so he shot one of them that measured four feet in height.[16]

In all probability, these, too, were chimpanzees. In 1900, British diplomat Sir Harry Johnston visited the Belgian Congo, where he claimed officials had photographs of a gorilla killed near Avakubi on the Ituri River.[17] The photographs have never surfaced, so the claim can't be substantiated. In any event, it's unlikely the photographs portrayed a gorilla.

While it may seem odd that no gorillas exist or ever seem to have existed in the forests of the Congo Basin, there is an explanation. As is true of other anthropoid apes, gorillas are not swimmers, and, while some Western groups do inhabit swamp forests, they avoid water over a few feet deep, although a few photos show ones wading chest high. The wide expanses of the Ubangi/Uele and Congo/Lualaba rivers likely served as impassable barriers. Similarly, the bonobos, often inaccurately called pygmy chimpanzees because of their smaller size than the common version, are confined to an area bounded by the Congo/Lualaba and Kasai rivers. Of course, it's possible that they have had no incentive to go elsewhere because of the abundant food resources available to them. From all accounts, bonobos are quite content where they are.[18] Gorillas do not seem to have ever inhabited the forests of the Guinea coast, and large rivers, plus a once-expansive Lake Chad, may have played a determining role here as well. Insufficient stands of appropriate forests and water barriers, such as the Nile River and Great Lakes, seem to have prohibited them from moving any farther east. In all likelihood, gorillas were never very dense on the land but instead, as today, lived in widely scattered groups throughout forested lands within the equatorial zone where necessary food and shelter could be found.

For millions of years, gorillas went about life largely undisturbed, affected only by slow climatic changes that altered the distribution of suitable forest habitats. They have no nonhuman competitors, although leopard attacks on immature gorillas have occasionally been reported. Early hominids like *Australopithecus* and *Homo erectus* never seem to have entered the equatorial forests. The first *Homo sapiens* to come their way were the ancestors of today's pygmy hunter-gatherers, such as the Baka, Babuti, and Bakoya in the lowland forests and Batwa in the upland forests. Just exactly when this happened is uncertain given the lack of archaeological information. That said, some unknown pygmy or pygmies must be accorded the distinction of having "discovered" gorillas many millennia ago.[19]

From what can be told, the largest apes and the smallest humans cohabited in relative peace, as the forests had more than enough room for both. And neither possessed anything the other wanted. Some 5,000 years ago, farmers began making an appearance. Gorillas actually can find more food in abandoned fields and secondary forests, and they're very fond of the pith located in the stems of bananas and plantains, staples of forest farmers. Sugarcane is also a temptation. At the outset, forest farming might well have increased gorilla population numbers, with positive environmental changes overcoming killings by humans during raids on plantations and sometimes for food, although the latter seems to have been rare in the past. In fact, people who live in the Virunga region and eastward from it traditionally have refrained from eating the flesh of gorillas, chimpanzees, baboons, and monkeys. One can only speculate as to why. Perhaps it has to do with their humanlike appearance. Or it could have had something to do with notions of clean and unclean. Ideas of sacredness do not seem to have been involved. And among whom did it originate? The Batwa would be the most likely suspects.

Beyond the shores of Africa, stories had been circulating for many centuries about a giant ape inhabiting the tropical forests of Africa, but it took until the mid-nineteenth century for the existence of gorillas to be confirmed. When this happened, things started to go very bad for them. Their discovery by Europeans led to purposeful hunting in order to obtain specimens for research and display, followed by attempts to capture infants for zoos and circuses, which usually entailed killing the mother and other members of the family, who will defend their young, often to the death. Trophy hunting also began to take a toll—another head to add to the collections of "great white hunters" and those seeking the latest curiosity (e.g., a hand serving as an ashtray) to show their avant-garde status. In addition, myths about the fierceness of gorillas circulated widely, helping to fuel their image as savage beasts and, by extension, the bravery of those who hunted them.

Now the forces affecting gorillas derive largely from growing human needs. More land is required to support ever-expanding populations, which in the twentieth century significantly reduced the habitats of the Mountain gorilla in particular. The demand for "bush meat" is accelerating, and gorillas sometimes are victimized or get trapped in snares set out for other game. The tropical forests hold vast wealth in the form of timber and minerals, and their exploitation leads to greater gorilla habitat fragmentation. And with more human contact, diseases such as pneumonia, measles, mumps, polio, the common cold, and gastrointestinal parasites have become a serious issue. Recently, Ebola hemorrhagic fever has ravaged several populations. Some illegal trade in live gorillas and gorilla parts also continues.

As a result of these impacts, all gorillas today are classified as "endangered" by the International Union for the Conservation of Nature (IUCN), with the Cross River population considered "critically endangered." And all of them come within the jurisdiction of the Convention on International Trade in Endangered Species (CITES) formalized in Washington, D.C., on March 3, 1973, to oversee a permit system for imports, exports, and re-imports. It's clear that left to their own devices, gorillas can't compete with us, as we have both demography and technology on our side. At the same time, we're their only hope of survival, as evidenced by growing conservation efforts and the African Wildlife Foundation declaring 2009 "The Year of the Gorilla" to bring their plight to a wider public. The question is, will these efforts be enough to offset the dark side of the gorilla/human equation? The answer won't be long in coming. And if the answer is "no," then the only safe place for gorillas will be zoos, where today most enjoy security and long lives if not independent ones.

NOTES

1. For detailed information regarding gorilla behavior and biology, see Alexander H. Harcourt and Kelley Stewart, *Gorilla Society: Conflict, Compromise and Cooperation between the Sexes* (Chicago: University of Chicago Press, 2007), and Andrea B. Taylor and Michele L. Goldsmith, eds., *Gorilla Biology: A Multidisciplinary Perspective* (Cambridge: Cambridge University Press, 2003). A good nonspecialist overview of great apes with excellent photographs can be found in Jennifer Lindsey, *The Great Apes* (New York: Metro Books, 1999).
2. *La Nature* 29 (1905): 129, cited in D. B. Willoughby, *All about Gorillas* (Cranbury NJ: A. S. Barnes and Co., 1979), 60.
3. Those readers interested in the complexities surrounding anthropoids can consult John G. Fleagle and Richard F. Kay, eds., *Anthropoid Origins* (New York: Plenum Press, 1994). A more concise and readable account is Raymond Corbey, "Negotiating the Ape-Human Boundary," in *Great Apes and Humans: The Ethics of Coexistence*, ed. Benjamin B. Beck et al. (New York: Springer, 2001), 163–77.
4. Russell L. Ciochon and Patricia A. Holroyd, "The Asian Origins of Anthropoidea Revisited," in Fleagle and Kay, *Anthropoid Origins*, 143–62.
5. The changes in taxonomy over time can be found in C. P. Groves, "A History of Gorilla Taxonomy," in Taylor and Goldsmith, *Gorilla Biology*, 15–34.
6. T. Ron, "The Majombe Forest in Cabinda: Conservation Efforts, 2000–2004," *Gorilla Journal* 30 (2005): 18–21.
7. Paul Matschie, "Uber einen Gorilla aus Deutsch-Ostafrika," *Sitzungsberichte der Gesellschaft Naturforschender Freunde Berlin* (1903): 253–59.
8. N. A. Sharp, "Notes on the Gorilla," *Proceedings of the Zoological Society of London* 97 (1927): 525; J. G. Allen, "Gorilla Hunting in Southern Nigeria," *Nigerian Field* 1 (1931): 5.
9. Estaben E. Sarmiento and John F. Oates, "The Cross River Gorillas: A Distinct Subspecies, *Gorilla gorilla diehli* Matchie 1904," *American Museum Novitates* 3304 (October 16, 2000): 2–55.
10. C. P. Groves, "A Note on the Affinities of the Ebo Forest Gorilla," *Gorilla Journal* 31 (2005): 19–21.
11. C. P. Groves, "Distribution and Place of Origin of the Gorilla," *Man* 6 n.s. (1971): 44–51.

12. Henry M. Stanley, *My Dark Companions and Their Strange Stories* (New York: Charles Scribner's Sons, 1893), 310–19.

13. Estaben E. Sarmiento, "Distribution, Taxonomy, Genetics, Ecology, and the Causal Links to Gorilla Survival: The Need to Develop Practical Knowledge for Gorilla Conservation," in Taylor and Goldsmith, *Gorilla Biology*, 433–34.

14. Martin Jenkins, "Evolution, Dispersal, and Discovery of the Great Apes," in *World Atlas of Great Apes and Their Conservation*, ed. Julian Caldecott and Lera Miles (Berkeley: University of California Press, 2005), 27.

15. Guy Burrows, *The Land of the Pigmies* (New York: Thomas Y. Crowell & Company, 1898), 239.

16. Sir Harry Johnston, *George Grenfell and the Congo*, vol. 1 (London: Hutchinson & Co., 1908), 344.

17. Thomas Alexander Barns, *The Wonderland of the Eastern Congo* (London: G. P. Putnam's Sons, 1922), xxiv.

18. A moving account of bonobos can be found in Vanessa Woods, *Bonobo Handshake: A Memoir of Love and Adventure in the Congo* (New York: Gotham Books, 2010).

19. The word "pygmy" has been called demeaning and even racist. In *"The Pygmies Were Our Compass": Bantu and Batwa in the History of West Central Africa, Early Times to 1900 C.E.* (Portsmouth, NH: Heinemann, 2003), Kairn Klieman substitutes "Batwa." But there's a problem in that this designates a particular grouping, and since nothing else has emerged as a valid replacement, "pygmy" continues to be the general description used for the short-statured tropical forest peoples of Africa.

Chapter Two

First Encounters

MYTHMAKING

Although determining an exact date has proved to be elusive, it does appear that sometime around the turn of the sixth and fifth centuries BCE, Hanno the Navigator, to distinguish him from other Carthaginians named Hanno, commanded a fleet of sixty to sixty-five galleys through the Pillars of Hercules (Strait of Gibraltar) destined for the western coast of Africa. He had instructions to reinforce existing colonies in what is now Morocco and to establish new ones at other promising locations wherever possible. To that end, the galleys reportedly carried 30,000 people and enough provisions to support them until well settled. Hanno also had been told to explore the lands to the south, with the intent of finding supplies of gold, a commodity essential to Carthage's prosperity, and to circumnavigate the unknown extent of the continent if this could be done. How far the galleys managed to go can only be guessed at because it's impossible to match precisely the place-names mentioned in a brief surviving *periplus* (Greek for "sailing itinerary" or "circumnavigation") with later ones. The original Punic version of the voyage has been lost to history. The fleet undoubtedly reached Cape Palmas, where the western African coast turns sharply eastward, although the journey may have covered a much greater distance. Toward the end of the *periplus*, there's mention of a high mountain translated as the "Chariot of the Gods." The two most likely candidates are Sierra Leone hugging the coast at an elevation of nearly 3,000 feet and Mount Kakulima in present-day Guinea at 2,920 feet. A more remote possibility is Mount Cameroon, rising to over 13,000 feet and easily visible from the sea. As stated in the *periplus*, three days of sailing beyond the mountain brought the expedition to a bay where the voyage culminated:

> And in this bay was an island . . . having a lake and in it another island filled with wild savages. The biggest number of them were females, with hairy bodies, which our . . . interpreters called "Gorillas." Chasing them, we could not catch any of the males, because all of them escaped by being able to climb steep cliffs and defending themselves with whatever was available; but we caught three females who bit and scratched their captors and they did not want to follow them. So we had to kill them and flayed them and we brought their skins to Carthage. We did not sail further, having run out of food provisions.[1]

Nothing, however, is mentioned about the return voyage.

Could these hairy beings have been the gorillas we know today? It's highly doubtful no matter where the encounter happened. Given their size, strength, and speed, capturing adult females is not very likely. Certainly, grabbing hold of one would have resulted in more than bites and scratches. And instead of running off when threatened, the adult males usually make a stand and sometimes charge to allow the females and young to get away. Throwing things at intruders, which we have to assume is meant by "defending themselves with whatever is available," is not their style, although they do sometimes rip up vegetation as a show of force. Furthermore, the creatures purportedly lived on an island. As noted in chapter 1, gorillas do not swim, and, therefore, they avoid deep, open expanses of water. The same is true of chimpanzees, and it's unlikely that the comments refer to baboons because the Carthaginians knew about them from North Africa. According to Pliny the Elder, the three skins disappeared after the First Punic War. Consequently, what Hanno and the others saw and where they saw it must remain mysteries.

The word "gorilla" and references to an animal like it disappeared from the written record for over a thousand years, although monkeys and apelike creatures, whether real or imagined, caught the fancy of writers and artists from classical times through the Renaissance. Missing, however, from their corpus of work is anything resembling an animal the size and shape of gorillas.[2] They remained beyond where sources of confirmatory information reached. That changed with Andrew Battell. A crewman aboard an English privateer, he was captured by the Portuguese off Brazil in the early 1590s and sent to their possession of São Paulo do Loanda on the other side of the Atlantic. Imprisoned and later employed on local trading ships, he remained in equatorial Africa for nearly twenty years. Upon returning to England, Battell told his story to a neighbor and friend, the Reverend Samuel Purchase. Near a place he called Yumbe (a Mayumbe presently exists along the Gabon coast), Battell said that there "are two kinds of monsters, which are common . . . and very dangerous":

> The greatest of these monsters is called *Pongo* [*Mpungu*] in their language and the lesser is called *Engeco*. This *Pongo* is in all proportions like a man, but that he is more like a giant in stature than a man; for he is very tall, and hath a man's face, hollow-eyed, with long hair upon his brows. His face and ears are without hair, and his hands also. His body is full of hair, but not very thick, and it is of a dunnish colour. He differeth not from a man but in his legs, for they have no calf. He goeth always upon his legs, and carry his hands clasped upon the nape of his neck when he goeth upon the ground. They sleep in the trees, and build shelters from the rain. They feed upon fruit they find in the woods and upon nuts, for they eat no kind of flesh.
>
> The people of the country, when they travel in the woods, make fires when they sleep at night. And in the morning, when they are gone, the *Pongoes* will come and sit about the fire till it goeth out, for they have no understanding to lay the wood together. They go many together, and kill many negroes that travel in the woods. Many times they fall upon the elephants, which come to feed where they be, and so beat them with their clubbed fists and pieces of wood that they will run roaring away from them.
>
> Those *Pongoes* are never taken alive, because they are so strong that ten men cannot hold one of them, but yet they take many of their young ones with poisoned arrows. The young Pongo hangeth on his mother's belly, with his hands clasped fast about her, so that when the country people kill any of the females, they take the young one which hangeth upon his mother. When they die among themselves, they cover the dead with great heaps of boughs and wood, which is commonly found in the forest.

In a marginal note, Purchase added,

> He [Battell] told me in a conference with him that one of these Pongos took a negro boy of his, which lived a month with them, for they hurt not those which they surprise unawares, except they look on them, which he [the boy] avoided. He said, their height was like a man's, but their bigness twice as great.[3]

While there's much nonsense in Battell's story, it's clear that *Pongo* equals "gorilla," even though none of the local languages contain this word. *Mpungu* probably refers to the Mpongwe inhabitants along the Gabon coast, and this could be where *pongo* comes from. Or it might be derived from *mwani-pongo*, the name for an Mpongwe lord. A 1686 compilation of observations about Africa by Flemish travelers also made reference to *pongo*.[4] As for *engeco*, this is close to the local word for "chimpanzee." But did Battell ever see a live gorilla? He never mentioned doing so, and his account reads more like the telling of tales heard.

In the 1774 edition of the multivolume *The Origin and Progress of Language*, James Burnet, better known as Lord Monboddo, included a letter written by the captain of a slave ship trading along the coast of western equatorial Africa. The captain had this to say about African apes:

> Of this animal there are three classes or species; the first and largest is, by the natives of Loango, Malemba, Calenda, and Congo, called or named Impungu. This wonderful and frightful production of nature walks upright like a man, is from 7 to 9 feet high when at maturity, thick in proportion, and amazingly strong; covered with longish hair, jet black over the body, but longer over the head; the face more like the human than the Chimpenza, but the complexion black; and has no tail. When this animal sees a negroe it mostly pursues and catches them; it sometimes kills them, and sometimes takes them by the hand and leads them along with him. Some that have made their escape say that this animal, when it goes to sleep, does not lie down, but leans against a tree. In this position, when the prisoner finds it asleep, he steals away the hand or arm softly from his, and so steals away quietly—sometimes discovered and retaken. It lives on the fruits and roots of the country, at the expense chiefly of the labor of the natives; and when it happened to be where there is no water, there is a tree with a juicy bark, which it strikes with its hand, bruises, and sucks the juice; and some of this tree it often carries with it when it travels, in case it should not find it or water by the way. And, indeed, I have heard them say that it can throw down a palm-tree by its amazing strength to come at the wine.[5]

The captain never saw the creature; rather, he related information obtained from informants. If both his and Battell's renditions are accounts of what they actually heard, then the informants knew little for a fact.[6] Another interpretation is that both men had had their legs pulled. Africans are known for playing jokes on one another and for spoofing unknowledgeable newcomers, most notably Europeans, with tall tales.

Stories from other unidentified sources about *pongo*s seem to have circulated widely. One told of them beating elephants with fists and clubs, whereas others reported on how a young African had been carried off for a year and then returning, of girls eight to ten years of age being taken into the trees for pleasure, and of a woman from Loango who spent three years with them, apparently none the worse for wear.[7]

The next traceable description belongs to T. Edward Bowdich. He interviewed slaves and merchants about conditions in the interior while aboard a trading vessel that ventured up the Gabon and Ogouwe rivers. Based on the information they provided, he told readers in his 1819 book *Mission from Cape Coast Castle to Ashantee*,

> The favourite and most extraordinary subject of our conversations on natural history, (which I introduce merely to excite inquiry) was the Ingena, compared with an Ourang-outan, but much exceeding it in size, being generally five feet

> high, and four across the shoulders; its paw was said to be even more disproportionate than its breadth, and one blow of it to be fatal; it is seen commonly by those who travel to Kaylee, lurking in the bush to destroy passengers, and feeding principally on the wild honey, which abounds. Their death is frequently accelerated by the silliness which characterizes most of their actions: observing men carry heavy burthens through the forest, they tear off the largest branches from the trees, and accumulating a weight (sometimes of elephants teeth) disproportionate even to their superior strength, emulously hurry with it from one part of the woods to another, with little or no cessation, until the fatigue, and the want of rest and nourishment, exhausts them. Amongst other of their action, reported without variation by men, women, and children of the Empoöngwa and Sheekan, is that of building a house in rude imitation of the natives, and sleeping outside or on the roof of it; and also carrying about their infant dead, closely pressed to them, until they drop away in putrefication.[8]

DISCOVERY

Differences and inaccuracies aside, these reports all agreed on one thing, namely, that somewhere in the depths of Africa's equatorial forests lurked a large and likely dangerous humanlike ape. But just what was it? The first concrete evidence to answer this question came in 1847, when a ship carrying Thomas Savage, an American medical missionary, was detained at the mouth of the Gabon River. While waiting for it to depart, Savage spent time with another missionary, J. Leighton Wilson, who possessed a skull much larger and of a different shape than any other known ape. By this time, the physical appearances of orangutans, gibbons, and chimpanzees were reasonably well described. Savage collected three additional crania, several pelvises, vertebrae, and an assortment of bones and sent them to Jeffries Wyman of the Boston Society of Natural History. The two men had previously collaborated on an article describing a chimpanzee, and they decided to follow the same procedure, whereby Savage discussed appearance and animal behavior while Wyman handled skeletal characteristics.[9] They named the find *Troglodytes gorilla*, the *Troglodytes* being widely used at the time for the chimpanzee. It's still used today, as we've seen, in *Pan troglodytes* for the common variety. Earlier, both Carl Linnaeus and Georges-Louis Leclerc, who's better known as Count de Buffon, had come up with *Homo troglodytes*, the former using it for the orangutan and the latter to cover the surmised existence of two African apes. In the vernacular, Buffon called the smaller one *Jocko* and the larger *Pongo*, following Battell. Others used Bowdich's *Ingena*, and thus, to avoid confusion, Savage and Wyman borrowed *gorilla* from translations of Hanno.

With tangible evidence at their disposal, the descriptions by Savage and Wyman of appearance and skeletal characteristics have pretty much stood the test of time. The assessment of gorilla behavior, however, still relied on Mpongwe informants. According to Savage, "the silly stories about their carrying off women from the native towns and vanquishing the elephants" could be put to rest. Nonetheless, he still bought into the notion of gorillas being ferocious creatures: "objects of terror to the natives, and are never encountered by them except on the defensive." Consequently, killing one was "considered an act of great skill and courage, and brings to the victor signal honor." As Savage related, the hunter awaited the animal's approach with "gun extended; if his aim is not sure he permits the animal to grasp the barrel, and as he carries it to his mouth (which is his habit) he fires; should the gun fail to go off, the barrel (that of an ordinary musket, which is thin) is crushed between his teeth, and the encounter soon proves fatal to the hunter."[10]

Additional specimens soon began to reach both museums and scientists in Europe and North America. The Royal College of Surgeons received its first of what would be many in 1851, and in 1858 the British Museum acquired a complete specimen preserved in a barrel of spirits. When dried out, it went on display at the Crystal Palace. One sent by Dr. Henry Ford to the Philadelphia Academy of Natural Sciences was accompanied by a description of physical features that confirmed most of what Savage and Wyman had presented. Ford repeated the gun story and said that informants told him that gorillas were "to some extent carnivorous," even hunting men for meat on occasion. They supposedly warred with the leopard and their "savage nature" was demonstrated by the impossibility of training the young. Based on such stories making the rounds, German animal artist Friedrich Specht drew a scene showing a gorilla subduing a leopard by grasping its paw in his left hand, with his right arm holding the head in place while he bites down on the back of the neck and a female with a young one looking on. The Mpongwe did, however, scoff at stories of gorillas building huts like humans and of using clubs to attack elephants.[11]

The arrival of new skeletal materials set taxonomists about the task of reclassifying anthropoid apes. Richard Owen, the reigning dean of British naturalists, kept the chimpanzee connection, initially using *Troglodytes savagei*; a little later he changed it to *T. gorilla* Savage.[12] The French looked at things a little differently. Isidore Geoffroy-Saint-Hilaire, the successor of his famous father Isidore as director of the Jardin des Plantes in Paris, considered the gorilla's anatomical characteristics different enough for genus status to be granted, and thus he came up with *Gorilla gina*.[13] M. Paul Gervais, who wrote the first real comparative study of the great apes, concurred. His monograph included five illustrations of gorillas—full frontal and side skeletal views, a

Plate 2.1. Illustration of the imagined "war" between gorillas and leopards. ***Source: Brehm's Life of Animals,*** **vol. 1,** ***Mammalia,*** **7.**

portrait of a skull, and two fairly realistic scenes, one of a mother and young and another of a gorilla grabbing a vine, with a smaller one in the background looking on.[14] Although classificatory differences continue to exist among experts, by later consensus the French won the day.

Still, to this point in time, other than Africans in a few locations, no one had yet seen gorillas in the flesh, much less studied them. One did reach England in 1855 but was exhibited as a chimpanzee in Wombwell's Travelling Menagerie, the largest and arguably the most famous of its type during their nineteenth-century heydays. Wombwell's remained in business until 1872, it and others giving way to the growing popularity of circuses. Named Jenny, highly regarded naturalist Charles Waterton viewed her on four separate occasions, and his description is quite consonant with a young gorilla. Jenny died suddenly in February 1856 after seven months with the menagerie. Waterton's restoration left no doubt about her having been a gorilla.[15] His recollections the last time they met indicate that although she had become accustomed to people, Jenny didn't enjoy her surroundings. What he said is also rather touching and not at all suggestive of a dangerous beast:

> Having mounted the steps which led up to the room, in order that I might take leave of her, Jenny put her arms around my neck; she "looked wistfully at me," and then we both exchanged soft kisses, to the evident surprise and amusement of all the lookers on.
>
> "Farewell,—poor little prisoner," said I. "I fear that this cold and gloomy atmosphere of ours will shorten thy days. Jenny shook her head, seemingly to say, there is nothing here to suit me. The little room is far too hot; the clothes which they force me to wear, are quite insupportable; whilst the food which they give me, is not like that upon which I used to feed, when I was healthy and free in my own native woods. With this we parted:—probably forever."[16]

Hearsay thus continued to be the source of information about the behavior of gorillas. Although the scientific community adhered mostly to the facts they could glean from skeletons and tried to figure out where gorillas fit into the scheme of life, more widely read authors emphasized the sensational aspects of prior accounts, sometimes adding additional flourishes to highlight the animal's reputed violent nature. For example, the popular natural history writer Philip Gosse told readers,

> This great ape makes the nearest approach of any brute-animal to the human form; it is fully equal to man in stature, but immensely more broad and muscular; while its strength is colossal. Though exclusively a fruit-eater, it is described as always manifesting an enraged enmity towards man; and no negro, even if furnished with firearms, will willingly enter into conflict with an adult male gorilla. He is said to be more than a match for the lion.
>
> The rivalry between the mighty ape and the elephant is curious, and leads to somewhat comic results. The old male is always armed with a stout stick when on the scout, and knows how to use it. The elephant has no intentional evil thoughts towards the gorilla, but unfortunately they love the same sorts of fruit. When the ape sees the elephant busy with his trunk among the trees, he instantly regards it as an infraction of the laws of property; and, dropping quietly down to the bough, he suddenly brings his club smartly down on the sensitive finger of the elephant's proboscis, and drives off the alarmed animal trumpeting shrilly with rage and pain.
>
> There must be something so wild and unearthly in the appearance of one of these apes, so demon-like in hideousness, in the solemn recesses of the dark primeval forest, that I might have told its story in the previous chapter. The terrors

Plate 2.2. Wombwell's Gorilla Jenny as portrayed in an 1877 illustration.

with which it is invested are, however, more than imaginary. The young athletic negroes, in their ivory hunts, well know the prowess of the gorilla. He does not, like the lion, sullenly retreat on seeing them, but swings himself rapidly down to the lower branches, courting the conflict, and clutches at the foremost of his enemies. The hideous aspect of his visage, his green eyes flashing with rage, is heightened by the thick and protruding brows being drawn spasmodically up and down, with the hair erect, causing a horrible and fiendish scowl. Weapons are torn from their possessors' grasp, gun-barrels bent and crushed in by the powerful hands and vice-like teeth of the enraged brute. More horrid, still, however, is the sudden and unexpected fate which is often inflicted by him.

> Two negroes will be walking through one of the woodland paths, unsuspicious of evil, when in an instant one misses his companion, or turns to see him drawn up in the air with a convulsed choking cry; and in a few minutes dropped to the ground a strangled corpse. The terrified survivor gazes up, and meets the grin and glare of the fiendish giant, who, watching his opportunity, had suddenly put down his immense hind-hand, caught the wretch by the neck with resistless power, and dropped him only when he ceased to struggle. Surely a horrified improvised gallows this![17]

Given the paucity of reliable facts, authors like Gosse could say just about anything in order to capture reader attention, which meant, of course, emphasizing the sensational. It would take a while before facts brought about change.

NOTES

1. A. N. Oikonomides and M. C. J. Miller, *Hanno the Carthaginian: Periplus, or Circumnavigation of Africa*, 3rd ed. (Chicago: ARES Publishers, 1995), 23.
2. For further information on early representations of apes, see Horst W. Janson, *Apes and Ape Lore in the Middle Ages and the Renaissance* (London: Warburg Institute and University of London, 1952), and Ramona Morris and Desmond Morris, *Men and Apes* (New York: McGraw-Hill, 1966).
3. E. G. Ravenstein, ed., *The Strange Adventures of Andrew Battell of Leigh in Angola and the Adjoining Regions* (London: Hakluyt Society, 1901), 54–55.
4. D'O. Dapper, *Description de L'Afrique* (Amsterdam: Chez Wolfgang, Waesberge, Boom & van Someren, 1686).
5. Quoted in W. Winwood Reade, *Savage Africa* (New York: Harper & Brothers, Publisher, 1864), 171–72.
6. Examples of misconceptions about gorillas even from Africans who knew of them can be found in Albert F. Jenks, "Bulu Knowledge of the Gorilla and Chimpanzee," *American Anthropologist* 13 (1911): 56–64.
7. Charles Waterton, *Essays on Natural History* (London: Longman, Brown, Green, Longmans, and Roberts, 1857), 40–42 .
8. T. Edward Bowdich, *Mission from Cape Coast Castle to Ashantee* (London: John Murray, 1819), 441.
9. Thomas S. Savage and Jeffries Wyman, "Notice of the External Characters and Habits of *Troglodytes gorilla*, a New Species of Orang from the Gaboon River; Osteology of the Same," *Boston Journal of Natural History* 5 (1847): 417–41.
10. Ibid., 423–25.
11. Henry A. Ford, "Communication," *Proceedings of the Academy of Natural Sciences of Philadelphia* 6 (1852): 30–33.
12. Richard Owen, "On the Gorilla (*Troglodytes gorilla* Sav.)," *Proceedings*, Zoological Society of London (1859): 1–23.

13. Isidore Geoffroy-Saint-Hilaire, "Descrition des Mammifères Nouveaux Connus de la Collection du Muséum d'Histoire Naturelle," Deuxième Supplement, *Archives de la Muséum Naturelle* 10 (1858–1861): 1–102.

14. M. Paul Gervais, *Histoire naturelle des mammifères avec l'indication de leurs moeurs, et de leurs rapports avec les arts, le commerce et l'agriculture* (Paris: L. Curmer, 1854).

15. Henry Scherren, *The Zoological Society of London: A Sketch of Its Foundation and Development* (London: Cassell and Company Limited, 1905), 172.

16. Waterton, *Essays on Natural History*, 66–67.

17. Philip H. Gosse, *The Romance of Natural History*, 1st series (London: John Nisbet and Company, 1861), 257–59.

Chapter Three

Hunting Gorillas

THE HUNT BEGINS

A young man named Paul Belloni Du Chaillu would attempt to set the gorilla record straight. Although what he had to say generated strong negative reactions from some influential contemporary authorities, his words served to influence the portrayal of gorillas in significant ways for many years to come. A little background information will help to set the stage for these two outcomes.

Paris is most often cited as Paul's birthplace, in either 1831 or 1835.[1] He himself at times claimed New York and New Orleans, and his headstone says "Louisiana." Although evidence can be found for both dates, the former seems more probable. Paul's mother is unknown and not likely married to his father, Charles-Alexis Du Chaillu. Given Paul's middle name of Belloni, she may have been Italian. Another possibility is that she was a Creole from Réunion, then known as Ile Bourbon, and unwed to Charles-Alexis.[2] He did work there as a trader during the early 1830s, which could make the island Paul's birthplace. Betting odds would seem to favor it. Records do not show the senior Du Chaillu having a son either at this time or during a return trip to Ile Bourbon in 1840. This, too, hints at Paul's illegitimacy and the likelihood that he was left in the care of someone else, perhaps his mother. Absence of information, especially from Paul, makes constructing his early years mostly guesswork.

Sometime in 1846, Charles-Alexis went to Gabon to assume the position of director of operations for a Le Havre–based trading firm. By 1848, he had moved into government service, as agent in charge of provisioning Gabon Colony and the French navy, stationed in what would soon be known as Libreville. Paul arrived on the scene later that same year. Whether he was invited or came of his own accord cannot be determined. Missionaries from South Carolina who ran a school took him in, their station at Baraka eight miles up the Gabon River serving as his new home. Natural history caught Paul's fancy, and he made short excursions into the interior to collect specimens of Gabon's rich flora and fauna. He claimed that the interest came from travels with his father on commercial ventures, and there may be some truth to this, although it's hard to say for sure. At some point, Paul also developed a fascination with the United States, and when an opportunity arose in 1852 to teach French at the Drew Seminary for Young Ladies in Carmel, New York, he eagerly took it. Despite—or maybe because of—his heavily French-accented English, Paul seems to have been a hit with the students. In addition, the time spent in Gabon earned him speaking engagements, and he also wrote several newspaper articles about the area. Paul Du Chaillu is a good example of how certain young men of modest means could elevate their positions in life by capitalizing on the growing thirst among the educated and wealthy to learn about distant lands and peoples. That the words came from someone who looked exotic—he was on the short side with dark, penetrating eyes and a somewhat swarthy complexion—probably didn't hurt his chances. And it's clear Paul possessed a charming, friendly personality that helped make him a desirable companion at social events.

Fortuitously, Du Chaillu brought some preserved birds and small mammals with him on the ship headed back to the United States. These he passed along to the Academy of Natural Sciences in Philadelphia, where the birds figured in several publications by John Cassin, a leading ornithologist. Communications with the Academy continued, and Cassin, along with a few other members, raised money so that Du Chaillu could return to Gabon to do some additional exploring and collecting, with the so-called fierce, untamable *gorilla* high on the agenda. The mid-nineteenth century was a heyday for amateur fieldworkers. Scientists of natural history needed them to supply specimens for their learned papers, and newly established museums competed with one another to fashion eye-catching and reputation-enhancing displays.[3] Finding someone like Du Chaillu, who had already been to a remote area and done some collecting, seemed like a godsend. For his part, Du Chaillu could hardly refuse an opportunity to return to familiar haunts and play the role of legitimate explorer.

Just prior to setting off in October 1855, Du Chaillu filed to become an American citizen. There's no record that he ever received his papers. The label "Franco-American" probably best captures his identity, as he called both countries home at various times.

Upon reaching Gabon, Du Chaillu stayed with the same missionary family as previously. From mid-1856 to late 1859, he claimed to have made three major excursions into the interior looking for gorillas and other interesting animal specimens. The first one targeted the Crystal Mountains via the Muni River, the second went through country in the vicinity of Cape Lopez, and the much longer third focused on the area in and around the Ogouwe and Fernan Vaz or Nkomi river valleys. It's unclear from Du Chaillu's account just how many gorillas he and his men were supposed to have killed. It could have been as many as a half dozen or more. He was always vague about this. In addition, two infants whose mothers had been killed came into his possession. Both died within days.

When he could, Du Chaillu sent specimens, especially birds, with descriptions to Cassin. He and the Academy, however, did not get any gorillas. One of the infants preserved in alcohol went to Jeffries Wyman in Boston, and several stuffed adults Du Chaillu kept for his own later uses.

Toward the end of 1859, Du Chaillu returned to the United States with the gorillas, an assortment of preserved birds, reptiles, and four-legged mammals, twenty of which he claimed as new species, in tow. He also brought along an array of skulls and bones. The size and diversity of the collection, plus over three years in Africa, rekindled his celebrity status. He lectured to both scientific and general audiences, hobnobbed with the upper crust, presented his findings to the Boston Society of Natural History[4] (it elected him a "corresponding member"), and began work on a book under contract with Harper and Brothers. The only sore point involved the Philadelphia Academy's refusal to cover expenses he felt owed him as an employee, which Du Chaillu wasn't. The money provided in 1855 came exclusively from subscriptions.

In early 1861 England beckoned. Here, too, Du Chaillu's lectures, with stuffed gorillas and skulls as props, attracted large, enthusiastic crowds, including at the Royal Geographical Society, which bestowed on him the honor of "fellow" and put the gorillas on exhibit. His book *Explorations and Adventures in Equatorial Africa* came out in May to favorable reviews. In many ways, the timing couldn't have been better. The publication in 1859 of Charles Darwin's *On the Origin of Species by Means of Natural Selection* had made evolution front-page news on almost a daily basis, with man's relationship to the apes being one of the most hotly contested issues. Propitiously, along came Du Chaillu with supposed firsthand accounts of gorillas in the wild. Were they ape-men, man-apes, or maybe the missing link? As a result, the book sold briskly since both supporters and detractors of evolution could find in it evidence to support their positions. For his part, Du Chaillu never took a side in the debate.

Quite suddenly, the mood changed. It started with accusations about some of the dates cited in the book not making sense, as they put Du Chaillu in different places at the same time. Another controversy involved bullet holes in one of the gorilla skins. It looked like the animal had been shot in the back, not in the chest, as claimed. The authenticity of several illustrations then came into question—they appeared to be slightly modified copies of already published ones. Most damning of all were assertions that Du Chaillu had done little or no exploring and hunting but instead had spent most of the time at the mission station taking in specimens brought by local Africans. The renowned German explorer of Africa Heinrich Barth claimed that the map Du Chaillu made of his journeys was pure fiction. Many of the places and features simply did not exist.[5] The real leader of the debunking effort, however, was John Edward Gray, keeper of the Zoological Collection at the British Museum, who used the popular gentleman's weekly *The Athenaeum* and the London *Times* to make his case. Virtually everything Du Chaillu said, especially about his natural history finds, Gray challenged. According to him, nary a specimen proved to be new to science.[6]

Du Chaillu wasn't without supporters. Such luminaries as famed anatomist/paleontologist Richard Owen, Roderick Murchison of the Royal Geographical Society, and publisher John Murray took his side. So too did the polymath Richard Francis Burton, who considered *Explorations* both interesting and authentic. A lively debate ensued, one that severely strained the relationship between longtime colleagues Gray and Owen. A letter from an American trader in Gabon who knew Du Chaillu strengthened Gray's hand. Among other things, the writer claimed that Du Chaillu had never traveled very far, much less tracked gorillas. Furthermore, he said the account contained numerous errors and inconsistencies, which, taken together, displayed either "mendacity or ignorance" on Du Chaillu's part.[7]

Renowned naturalist and taxidermist Charles Waterton considered the book "a disgrace to zoology," and Abraham Dee Bartlett, who became superintendent of the London Zoo in 1859, remembered questioning Du Chaillu in 1861 over the condition of a gorilla skin he wished to have stuffed. As Bartlett related the incident,

> I called M. Du Chaillu's attention to the face of the animal, which, I told him was not in perfect condition, having lost a great part of the epidermis [outer skin]. In reply he, M. Du Chaillu, assured me that it was quite perfect, remarking at the same time, that the epidermis on the face was quite black, and that the face of the skin being black was proof of its perfectness.
>
> I, however, then and there convinced him that the blackness of the face was due to its having been painted black; finding I had detected what had been done, he at once admitted that he did paint it at the time he exhibited it in New York.
>
> The question that arose in my mind upon making this discovery was, did M. Du Chaillu kill the gorilla and skin and preserve it? If so, he must recollect that the epidermis came off; supposing he did forget this, he must have been after-

> wards reminded of the fact when he had to paint the face to represent its natural condition. These facts (to which I had a witness) led me to doubt the truthfulness of M. Du Chaillu's statement, and it occurred to me that he was not aware of the state of the skin, and probably had not prepared it himself.[8]

Despite his skepticism, Bartlett still followed Du Chaillu's story line, remarking about the gorilla, "Its power must be prodigious; its fierce and brutal aspects render it at once the most repulsive of brutes."[9]

As an interesting aside, Bartlett claimed that he had examined one of Waterton's specimens called "Martin Luther" possessed of donkey ears and found it to be a gorilla, which supposedly reached England when Waterton was a young man. If true, the gorilla would predate by many years Savage and Wyman's documentation. It is, however, not true. You can find the image online with the true title of "Martin Luther after His Fall." It's clearly one of Waterton's many satirical caricatures, in this instance based on some kind of primate with horns added.

For the most part, Du Chaillu again kept above the fray. One time, though, he did lose his temper—not over gorillas but rather when someone in the audience of a meeting at the Ethnological Society in London questioned the accuracy of his description of an African harp. In a rage, Du Chaillu rushed at the man, fists clenched. Stopping short, he instead spat in his face. Later, Du Chaillu publicly apologized to the Society in a letter to the London *Times*.

The issue of Du Chaillu's reliability would not be settled at the time. More to the point is that his words came to be what most people knew about gorillas. Readers of *Explorations* could find two sources of information. One involved the hunts, the descriptions of which contained a spate of graphic imagery designed to make the gorilla look ferocious. It's likely that the editors at Harper were involved in this, hoping to increase sales of the book. And there's some evidence to suggest the hand of a ghostwriter at work, most likely the corresponding secretary of the Boston Society of Natural History, Dr. S. Kneeland.[10] Nonetheless, for what follows, Du Chaillu will be taken as author.

The first hunt set the tone. One day when crawling through a dense tangle of forest, Du Chaillu claimed to have heard a loud bark. A few moments later, he reported seeing a large male gorilla standing dead ahead, an animal "with immense body, huge chest, and great muscular arms, with fiercely-glaring large deep gray eyes, and a hellish expression of face . . . like some nightmare vision." He said it roared repeatedly and beat its chest with fists, producing an image of "some hellish dream creature—a being of that order half-man half-beast, which we find pictured by old artists in some representations of infernal regions." At a distance of about six yards, Du Chaillu fired, and down the malevolent creature went.[11]

Subsequent encounters reinforced the images. When another large male loomed about twenty yards away, Du Chaillu said,

> We stood therefore in silence, guns in hand. The gorilla looked at us for a minute or so out of his evil gray eyes, then beat his breast with his gigantic arms, gave another howl of defiance, and advanced upon us.
>
> Again he stopped, now not more than fifteen yards away. . . . Then again an advance upon us. Now he was not twelve yards off. I could see plainly the ferocious face of this monstrous ape. It was working with rage; his huge teeth ground against each other so that we could hear the sound; the skin of the forehead was moved rapidly back and forth, and gave a truly devilish expression to the hideous face: once more he gave out a roar which seemed to shake the woods like thunder, and, looking us in the eyes and beating his breast, advanced again.[12]

This time, three musket balls supposedly felled the attacker at eight yards.

To give the ferocity theme a further boost, Du Chaillu related what happened to one of his hunters. Hurrying on at the sound of "terrific roars," he said they found

> the poor brave fellow . . . lying on the ground in a pool of his own blood. . . . His bowels were protruding through the lacerated abdomen. Beside him lay his gun. The stock was broken, and the barrel was bent and flattened. It bore plainly the mark of the gorilla's teeth.[13]

A drawing showed the gorilla standing over his victim with bent gun barrel in his hands. It would be republished many times over.

Plate 3.1. The depiction of a "Hunter Killed by a Gorilla" in Paul Du Chaillu's *Explorations and Adventures in Tropical Africa.*

Du Chaillu had problems with the demeanor of gorillas. He described it as a "sickening realization," without making clear what made him sick—an animal with such beastly characteristics looking so human or the act of killing them. His portrayal of an encounter with a mother and infant suggests that both were involved and created a sense of guilt in him. He said he considered not shooting the mother because the scene looked so "pretty and touching," but one of his men did shoot, killing her. According to Du Chaillu,

> As soon as he [the infant] saw his mother he crawled to her and threw himself on her breast. He did not find the accustomed nourishment, and I saw that he perceived something was the matter with the old one. He crawled over her body, smelt at it, and gave utterance, from time to time, to a plaintive cry, "Hoo, hoo, hoo," which touched my heart.[14]

In what can only have been an attempt to establish Du Chaillu as a legitimate scientist, about two-thirds along the way the narrative in *Explorations* suddenly changed to a more analytical style. One chapter provided information on the history of gorilla discovery going back to Hanno and related various legends about the animal's behavior. He said that his findings disproved such widely held notions as gorillas waiting in trees to grab people, attacking elephants and beating them "to death with sticks," carrying off women for their pleasure, and building tree houses to sit on.[15] In another chapter, replete with diagrams, readers were provided the latest details about ape comparative anatomy. After this, the thrilling narrative resumed.

Lending credence to Du Chaillu's critics is that everything reported in the book about gorillas had already appeared elsewhere. He simply repackaged the information into an adventure story containing new discoveries, meetings with so-called cannibals, and many escapes from danger. As a result, the narrative contains some glaring errors that wouldn't likely have shown up had adult gorillas been seen close-up. For example, their eyes are dark brown, not gray, and they slap their chests with cupped hands instead of beating them with fists.

Spurred on by the controversy William Winwood Reade, a young adventurer and aspiring journalist, sailed for Gabon in 1859 in search of the truth. Had Du Chaillu actually hunted and killed gorillas or not? Although he, himself, never saw a gorilla in the wild during five months in western equatorial Africa, Reade did talk to locals who had done so, and he had seen gorilla tracks and also some young ones in captivity. As a result, he said,

> I can assert that it travels on all fours; for I have seen the tracks of its four feet, over and over again. I can assert that it runs away from man, for I have been near enough to hear one running away from me; and I can assert that the young gorilla is as docile as the young chimpanzee in a state of captivity, for I have seen both of them in a state of captivity.[16]

As for Du Chaillu, Reade concluded that he "has written much of the gorilla which is true, but which is not new; and little which is new, but which is very far from being true."[17] He also noted gorillas do not "attack a man without provocation."[18]

Thomas H. Huxley, known as "Darwin's Bulldog" for his vigorous support of evolution, went a little further in his essay "On the Natural History of the Man-Like Apes," noting,

> If I have abstained from quoting M. Du Chaillu's work, then, it is not because I discern any inherent improbability in his assertions respecting the man-like Apes; nor from any wish to throw suspicion on his veracity; but because, in my opinion, so long as his narrative remains in its present state of unexplained and apparently inexplicable confusion, it has no claim to original authority respecting any subject whatsoever.
>
> It may be truth, but it is not evidence.[19]

Huxley wisely cautioned against believing any statements regarding both gorillas and chimpanzees because so little was known about them compared to gibbons and orangutans.

Although he probably didn't have to, as *Explorations* outshined its critics and remained the standard source on the gorilla for many years to come, Du Chaillu set off again for Gabon in 1863 to redeem his reputation by doing further exploration and collecting. Finances came from book sales, specimens bought by the British Museum, and donations by friends. To validate his travels, Du Chaillu brought along an array of instruments. Unfortunately, they were lost when the canoe taking them to shore capsized in heavy surf, and he had to wait months for replacements to arrive before traveling any great distance into the interior.

This time, Du Chaillu wanted to concentrate on studying gorillas in the wild. He thought that Africans had provided enough specimens for research and museum exhibitions to obviate the need to kill more. As it turned out, his trips yielded only a few brief opportunities to observe gorillas. On one occasion, locals did bring him three live ones: an adult female, her infant, and a young male. Looking at them, Du Chaillu thought, "what would I not give to have the group in London for a few days!"[20] No one could question him then. The female, however, was badly injured in the struggle to capture her and died the next day. The infant lasted only three days longer, and the male starved to death en route to England. The provisions provided him gave out, and he couldn't be sustained on human food. Because of the paucity of encounters, Du Chaillu's *A Journey to Ashango-Land* added very little new information about gorillas beyond confirming that they lived together in groups of males, females, and young. Previously, he had described them as lacking a "gregarious" nature.

This trip turned out to be Du Chaillu's last to Africa. The return to the coast nearly cost him his life on more than one occasion, and he vowed never to go back. Du Chaillu did, however, continue to write about his experiences, producing five books that reshaped the two originals into stories designed for young boys. They proved immensely popular and established him as "the gorilla hunter" for coming generations. Their popularity led to reissues in 1889. In the final volume, *The Country of Dwarfs*, Du Chaillu bade farewell to Africa, saying that he will next take readers to Norway, Sweden, and Lapland. By its publication in 1871, he, indeed, had already switched his attention to Scandinavia. Travels there for nearly a decade led him to write *Land of the Midnight Sun* and *Viking Age*, a 600-page two-volume tome in which he tried to show the Vikings as progenitors of the English. Then, approaching seventy years of age, Du Chaillu returned to writing about Africa for young people, producing *The World of the Great Forest*, *King Mombo*, and *In African Forest and Jungle*.

Plate 3.2. Paul Du Chaillu in his later years. *Source:* Clodd, *Memories*.

Paul Belloni Du Chaillu died in St. Petersburg, Russia, on April 30, 1903, never having married but still very much a part of elite society. One of his best friends was New York City appeals court judge and president of the American Geographical Society, Charles P. Daly. Whatever may have tainted his reputation during the controversy over *Explorations* had long since disappeared from public view, as evidenced by his being included with the likes of Lewis and Clark, Zebulon Pike, and Henry Morton Stanley in the series *Men of Achievement*.[21]

THE HUNT CONTINUES

In March and April 1862, Richard F. Burton tried his hand at finding gorillas while stationed at Fernando Po (now Bioko) as British consul overseeing the bights of Benin and Biafra. Capturing a live one or even getting a dead one to send to Great Britain would add another feather to his cap of accomplishments. With a copy of *Explorations and Adventures in Equatorial Africa* in hand, Burton made two separate attempts to locate gorillas during a trip to Gabon, but he never saw a single one, only evidence of some having been nearby.[22] A specimen sent to him by a local hunter turned out to be a large chimpanzee when examined at the British Museum.

An interesting account of a possible gorilla encounter appears in a letter by Dr. David Livingstone to his sister Agnes in late 1870. He wasn't out to find them but rather was searching for the source of the Nile in the vicinity of the Lualaba River in the eastern Congo. According to him,

> She sits crouching eighteen inches high, and is the most intelligent and least mischievous of all the monkeys I have seen. She holds out her hand to be lifted and carried, and if refused makes her face as in a bitter human weeping, and wrings her hands quite humanly, sometimes adding a foot or third hand to make her appeal more touching. . . . She knew me at once as a friend, and when plagued by anyone always placed her back to me for safety, came and sat down on my mat, decently made a nest of grass and leaves, and covered herself with the mat to sleep. I cannot take her with me, though I fear that she will die before I return, from people plaguing her. Her fine long black hair was beautiful when tended by her mother, who was killed. I am mobbed enough alone; two sokos—she and I—would not have got breath.[23]

This sounds like a baby gorilla, and, if so, Livingstone would have been the first European to see an Eastern version, given the location. *Soko*, however, more commonly meant "chimpanzee" in this part of Africa at the time.

Paul Du Chaillu's assessment that enough gorilla specimens had been obtained to satisfy demand turned out to be quite wrong. Scientists and museums wanted a regular supply of new ones to study and exhibit, and African hunters were only too happy to comply with their wishes. An intermediary of record was Robert M. Nassau, a medical missionary who lived at a station along the Ogouwe River valley from 1874 to 1891. He supplied Dr. Thomas G. Morton in Philadelphia with several specimens, one being a complete carcass preserved in rum that eventually went on display at the Pennsylvania Hospital Museum. During leave in the United States in 1891, Nassau presented Morton with the first complete and perfectly preserved brain for study. It belonged to a young male that died from an attack of driver ants after being captured by Gabonese hunters.[24] In 1900, a skinned and stuffed gorilla from Cameroon shot by German trader H. Paschen went on exhibit in Hamburg, Germany. Lionel Walter, the Second Baron Rothschild, then purchased it for the huge private collection he had assembled in Tring, Hertfordshire. After his death in 1937, the collection was gifted to the nation and is now the Natural History Museum at Tring. The gorilla is still there.

Specimens, whether stuffed or preserved, weren't enough. People clamored to see live gorillas. Never one to miss an opportunity to make the public gasp, P. T. Barnum offered to pay $20,000 for a healthy adult to put on display at the Barnum and Van Amburgh Museum and Menagerie. Consequently, when a letter reached him in the summer of 1867 about the arrival in New York City of a fierce gorilla, he jumped at the chance. On inspection, it turned out to be a large baboon, as confirmed by his good friend Paul Du Chaillu. They shared a hearty laugh over the letter's erroneous description of the beast.[25]

Continental Europe's first live gorilla went to the Berlin Aquarium in 1876, after making a brief stop in Liverpool. Initially called Pongo, it soon became known as "Falkenstein's gorilla," after Julius Falkenstein, the doctor who acquired the juvenile male in Gabon from a Portuguese trader while accompanying a German expedition led by the multitalented Eduard Pechuel-Loesche. Pongo created quite a stir at the Aquarium with his liking of wine and fondness for children, a trait often shown by gorillas and other great apes. He also was the first gorilla to be studied for behavior and general comportment. In 1877, the youngster tickled the fancy of crowds in London, especially when blowing puffs of cigarette smoke. On the way back to Berlin, Pongo died from bowel inflammation after spending a total of only thirteen months in Europe.[26]

By the 1890s, the Berlin Aquarium had acquired four additional gorillas, of which only one seems to have lived for more than a few months. In the early 1880s, the Dresden Zoo thought it had a gorilla, but the ape turned out to be a chimpanzee instead. The Jardin des Plantes in Paris was more fortunate when it obtained a three- to four-year-old male in 1883. Its withdrawn personality didn't, however, delight visitors, and like the others before him, the young guy lived for only a short time. The several that reached England fared no better. One in the mid-1870s died not long after arrival, and another taken in by the Zoological Society of London in 1883 lasted only two months. The same fate awaited the Society's second charge in 1897.[27] Being fed such things as cheese, mutton, and even beer hardly helped their cause. After two more failures, the London Zoo decided to give up on the idea of keeping gorillas. Only much later would it try again. An infant that in 1897 became the first live gorilla to reach the United States survived for only five days. "Pussi" had a somewhat better fate. Purchased in Liverpool in September 1897 at about four years of age by the Breslau Zoo, she survived until October 1904, making her the longest-lived gorilla in captivity up until then.[28]

In a brief overview of what was known about gorillas at the close of the nineteenth century, Sir Arthur Keith singled out temperament rather than climate as the likely cause for their early demise. He described them as being "fierce, intolerant of bonds, and lacking the docility of the easily-confined chimpanzee."[29] Melancholy and morose became other common descriptors of gorilla personality.

Such characterizations did not apply to Pongo or to Seraphine, a young female kept for four months in 1859 by R. B. Walker, who owned a mercantile operation in Gabon. He described her as being "perfectly tame, docile, and tractable":

> Not only was she on perfectly good terms with all grown-up people in and about the factory [trade station], but she was exceedingly attached to her keeper . . . whom she could not bear to be out of her sight, but regularly accompanied him about the factory and in his walks in the town and neighbourhood. . . . She allowed herself to be clothed, seeming to like it; and actually went to breakfast with a friend of mine . . . upon which occasion she conducted herself to the admiration of everybody.[30]

Seraphine, too, had a short life, dying, Walker said, from "dysentery and chagrin, the latter caused by her keeper being prevented by his other occupation from paying her so much attention as she had been in the habit of receiving."

As noted previously, if anything came through loud and clear in Du Chaillu's description of gorillas, it was their ferocity, so great that he put them at the top of the list of Africa's most dangerous game animals. Richard Owen followed suit, commenting, "Negroes, when stealing through shades of the tropical forest, become sometimes aware of the proximity of one of these frightfully formidable apes by the sudden disappearance of one of their companions, who is hoisted up into the tree, uttering, perhaps, a short croaking cry. In a few minutes he falls to the ground a strangled corpse."[31]

Artists fell prey. Immanuel Frémiet, noted for his statue of Joan of Arc in Paris, sculpted an image in 1859 called *Gorille enlevant une Femme*, which

Plate 3.3. A replica of *Gorille enlevant une Femme* by Immanuel Frémiet. *Source:* American Museum of Natural History, image 36832.

shows a snarling gorilla carrying off a nearly naked European woman under his right arm and holding a large rock in his left hand ready to throw at would-be pursuers. Outrage caused the piece to be banned and then destroyed, but Frémiet carved a new one in 1887. A statuette by William Umlauff of Hamburg purportedly depicts a real event in which three African hunters are being mauled to death by a gorilla.[32]

Until recently, African art depicting gorillas was limited to those few peoples who knew of them from experience. One such group is the Kwele, who live along the borderlands of the Republic of the Congo, Gabon, and Cameroon. They are known for their wood-carved ceremonial face masks, with the gorilla ones, called *gon*, signifying danger. While some masks do, indeed, convey an image of fierce-looking beasts, others display more humanlike features.[33]

Plate 3.4. A Kwele gorilla mask (*gon*) displaying ferocity with its fanglike canines. *Source:* Leon Siroto and Kathleen Berrin, *East of the Atlantic West of the Congo: Art from Equatorial Africa* (Seattle: University of Washington Press, 1995).

The various representations in words, on canvas, in stone, and in wood not only shocked and excited audiences, but they also enhanced the image for bravery of anyone seeking after gorillas. Consequently, a new adversary entered into their lives, the big-game, or white, hunter.

Actually, the first hunters after Du Chaillu were three fictional characters in the appropriately titled *The Gorilla Hunters* by Robert M. Ballantyne, one of the most prolific novelists of the second half of the nineteenth century. Published in 1861, the book's gorilla "facts" came directly from Du Chaillu's *Explorations*. The animals were described as "hideous" to look at, roared as no others, used fists to beat their chests, and bent rifle barrels with ease. The story line contains the "pretty and touching" episode of mother and infant,

Plate 3.5. A beautifully carved Kwele gorilla mask (*gon*) with stylized canines and sagittal crest. *Source:* Mestach Collection.

except this time one of the white hunters reacted quickly and struck the gun of the native hunter taking aim, causing the shot to miss. When queried as to why, the hunter explained, "to prevent you from committing murder. . . . Have you no feelings of natural pity or tenderness, that you could coolly aim at such a loving pair as that?" In total, Ballantyne had his characters killing thirty-seven gorillas, all for the sake of science. That he meant this sarcastically is revealed by one of the men remarking, "humph! I suspect that a good deal of wickedness is perpetrated under the wing of science."

Ballantyne gave one of the three protagonists the nickname "Gorilla" because of his much larger size than the other two. It's hard to say if this is the first such usage to appear in print, but given his popularity, he must have been influential in establishing gorilla as an epitaph for large, burly men. It was meant as a term of endearment and not *Webster's Dictionary*'s slang definitions of "ugly brute of a man" or "a thief who resorts to violence."

CAPTURED BY A GORILLA

The notion of gorillas capturing humans, especially males lying in wait for females, comes from stories that Africans told Europeans. That they were willing to believe is hardly surprising, given the many tales about the sexual potency and desires of animals extant in the literature since at least classical times.[34] By the eighteenth century, orangutans had risen to the top of the sex-crazed ape hierarchy, only to be replaced by gorillas once their existence was confirmed. Despite the notion having been debunked by both Savage and Du Chaillu, it persisted, as is illustrated in an 1867 publication called "Captured by a Gorilla." The author, the Reverend John Onslow (a pseudonym), claimed to have witnessed the event and to have sent his narrative to David Livingstone, asking him to publish it so "as to prevent any suspicion of deception or error." According to Onslow, a German dealer in precious stones named Reuben Haas called at his mission station in Gabon, bringing with him his eighteen-year-old daughter Leah, whom Onslow described as "the perfection of loveliness." After several months' stay at the station, the three of them journeyed inland, Reuben and Leah in search of diamonds and Onslow on missionary rounds. Some days later, they reached gorilla country. Even though she had killed a leopard, Leah had been warned by locals not to go because a gorilla would likely "carry her away, away in dark forest." She ignored the advice, especially since it meant staying with the Fang, considered cannibals based on Du Chaillu's portrayal. In reality, no danger would have ever existed of Leah's winding up in the cooking pot, but the story has her detesting the Fang so much that she couldn't bear being around them.

One morning during a search for minerals, Leah became separated from the others. Suddenly, a "succession of demoniacal screams, barks, and roars, rang through the forest," followed by Leah shouting out, "Father! father! Help! help!" A gorilla had grabbed her. Leah's father quickly organized a rescue party. They found a fragment of her jacket and a spot of blood along a trail. Following it, a gorilla appeared as the sun began to set. Onslow said, "Pen and pencil combined, even by such skilful hands as Du Chaillu's, cannot paint the gorilla as he is in his native forest." Instead,

> To properly appreciate the gorilla as the most hideous of Nature's productions you must look on him with your eyes. You must pursue him and bring him to bay. You must look into the depths of his lightening, devilish gray eyes; you must watch his immense lips, heavy as the Bengal tiger's, twitching back to show his terrible teeth; you must see the hard, iron-like skin of his forehead jerking backward and forward into corrugated folds like ropes; you must witness the peculiar throwing forward of the head as he utters his roar; you must behold those immense muscles that lie in massive folds across gigantic breast; you must

> see those titanic arms, like great beams, as it were, of knotted iron cable, ending in hands that look like huge chunks of iron, and with which he continually beats upon that tremendous breast.

They shot him dead, only to discover that he was not the one who had taken Leah. The search thus continued through lands in which the natives said no animal lived, save the gorilla. "He reigns lord of his wild, forest-domain; not even the elephant daring to intrude upon his territory," according to them.

Two days later, the party came across fresh tracks, and soon the sought-after gorilla appeared, with Leah, nearly naked but alive, lying beneath his foot. He roared and beat his chest, seeming "more like a fiend from the bottomless pit than a beast of the forest." When within three yards, the two men fired, killing the gorilla instantly. As Onslow put it, "The monster's dying yell, groan, or shriek, whichever it might be called, was so human, and yet so unearthly, so demonic that I can hardly yet persuade myself that the gorilla is a mere animal." Leah then related the story of her capture. She had been caught unaware of the ape's presence. He grabbed her rifle "and giving it a ferocious snap between his teeth, doubled it up and threw it away." She screamed but was grabbed by the throat, rendering her insensible. On awakening, Leah found herself in the gorilla's lair. She fled, only to be seized again. At the sound of another gorilla, he grabbed her and took off. The other gorilla overtook them, leading to a fierce fight:

> The huge beasts tore and bit each other and struck each other with their fists, blows that would have felled an elephant. They wrestled and rolled and floundered, breaking down young trees, at least six or eight inches in diameter. They always struck at the head and downwards, not like men, but like dogs swim, and in biting they would fasten their teeth and tug several times to pull the piece out, if possible, and then let go and repeat the same manoeuvre; all the while uttering the most hideous barks and growls.

Leah ran but didn't get far before being taken once again by her victorious captor. She claimed to have fought desperately, hoping as the struggle progressed that he might kill her so as to spare further horror. That's when Mr. Haas and Onslow appeared on the scene to save the day. Onslow ends by saying, "My narrative is done, and I give it, my dear friend [Livingstone], to the world through your hands, pledging you my Christian faith that it is true, every whit." There's no evidence that Livingstone ever received the manuscript, much less acted on it.

Returning to the world of nonfiction, the German Hugo von Koppenfels appears to have been the first European hunter, not counting Du Chaillu, who actually tracked and killed gorillas in the wild. He claimed to have brought down two of them in 1874–1875, the second said to have been a large male that reportedly advanced toward him roaring and beating his chest. According to his account, as he raised his rifle, the gorilla's

> roaring took on more of a barking sound; he beat his chest quicker, the shaggy hair on his head raised itself with a vibrating motion, and it seemed that my terrible opponent was going to attack me. If I had retreated in time, I am fully convinced that the Gorilla would not have approached me, but such was not my intention. Mastering my agitation, I took steady aim at his heart, and pulled the trigger. The animal jumped high up, and spreading his arms, fell on his face. He had seized in falling, a liana two inches in circumference, and so powerful was his grasp that he tore it down along with dry and green branches from the tree.[35]

This event and others reported by von Koppenfels led natural historian and director of the Hamburg Zoological Gardens Alfred Brehm to conclude that the time had come to "do away with a great deal of the terror with which he [the gorilla] has inspired us."[36] Still, though, Brehm ranked the silverback (the name comes from the gray hairs that grow on the back of adult males) as "The king of the African jungle," remarking in an illustration caption in his magnum opus on mammals, "Unarmed Man, the Leopard and the Crocodile are no match for this formidable creature, before which even the Lion might tremble."[37]

The most prolific real gorilla hunter of all time has to have been Fred Merfield. His exploits are recorded in *Gorillas Were My Neighbours*. In 1910, he left a boring job in London to work at a plantation in Kameroon, then a German colony. While there, he developed an interest in gorillas and decided to become a professional hunter. These plans had to be put on hold until the end of World War I, as Merfield joined Great Britain's West African Freedom Force. Afterward, he settled in Ebolowa District in the French portion of newly divided Cameroon and began his thirty-year hunting career. Merfield collected numerous animals, mostly for museums and scientific research, both of which continued to have unlimited appetites for gorillas in the 1920s and 1930s. Some of these he shot, others he got from local hunters, the total, according to him, amounting to 115. Most came from Mendjim country, noted for its abundance of gorillas and the skill of the Mendjim hunters, who used spears and crossbows with poisoned arrows to make kills meant for the table, they being the only people in this part of Africa who regularly ate them.

Merfield didn't hunt for zoos and thus refrained from killing adult gorillas to secure juveniles. He did, though, often find himself in possession of ones who had lost their parents. These he tried to find homes for, but all died in transit. Merfield also shied away from safari hunter types, except when money ran short. One whom he agreed to guide was renowned big-game hunter Major P. H. G. Powell-Cotton, who wanted a gorilla specimen for his museum in Kent. The hunt proved to be successful, and, in addition, the ma-

Plate 3.6. A nineteenth-century representation of the gorilla as "King of the Jungle." ***Source: Brehm's Life of Animals*****, vol. 1,** ***Mammalia*****, 3.**

jor shot a film of two orphaned babies to show at his museum. Unfortunately, both died shortly thereafter. During the expedition, Merfield learned from the major how to preserve skins and skeletons more effectively for overseas shipment, and Powell-Cotton contacted museums and scientific institutions to drum up business for him. He came back for a second visit of eighteen months' duration and obtained a number of animals, both alive and dead, to send to zoos and museums.

Hunting of this kind had come under increasing fire from conservationists, and consequently during October 1953 in Bukavu, located at the south end of Lake Kivu, the International Conference for the Protection of African Fauna specified that gorillas could not be hunted, killed, or captured except by license, with such licenses to be granted only under special circumstances.

Any other trade henceforth would be considered illicit, although, as is usual with such pronouncements, it had little immediate affect.

In addition to hunting gorillas, Merfield spent time recording their behavior, leading to some good descriptions of nest making and family organization, especially from a troop he called his "family."[38] Merfield also related several instances of traditional gorilla "roundups," which took place when the animals became a nuisance by raiding plantations. At the call, people came from miles around to put up stakes and make fences. While women sang, drummed, and shook bells and rattles, the men stood by with spears, cutlasses, and old guns awaiting the animals to be driven into fenced clearings where the assault began. Merfield detested roundups because of the suffering involved. Here's how he described the end of one them:

> The death of the Old Man was one of the most terrible things I had ever seen. He tried to get out on the opposite side and we were attracted there by the yelling of the spearmen. Running around, we found the poor beast sitting in the undergrowth, as I had seen gorillas do so often before in more happy circumstances, and there were at least a dozen spears projecting from his chest, back and sides. Making no attempt to retaliate, he was just sitting there, rocking to and fro as more spears were thrust home at close quarters; his mouth was wide open, crying shame on his tormentors. The gorilla's anguish was too much to bear for me and I put up my rifle to shoot, caring nothing for the consequence [he'd been warned not to interfere], but before I could fire a native stepped forward between me and the stricken beast. Three or four more spears hit the gorilla and then another man, armed with a heavy club leaned forward, and slowly, methodically, began clubbing him to death.[39]

Despite these events, Merfield didn't think gorillas in Cameroon faced extinction. He felt that they could retreat to unpopulated mountainous areas, and within Mendjim country, he thought a cyclic balance had been reached:

> Gorillas were regularly hunted for food, but they continued to raid the plantations until the villagers, roused at last from their usual lethargy, killed comparatively large numbers of them and drove away the rest. For a short time the plantations were secure, then more gorillas moved in from other parts of the forest and it all began again. Thus, both gorilla and human populations were maintained at the level appointed by that delicate, complex and inexorable system of laws which we call the Balance of Nature.[40]

As noted, gorillas still exist in southern Cameroon, but there's no longer a balance, as people have become far more numerous since Merfield's days, and their demands for space and resources have grown accordingly. The result, of course, is the presence of fewer gorillas.

A refuge for wounded and abandoned gorillas, indeed for all animals, existed at Dr. Albert Schweitzer's Lambaréné hospital and grounds in Gabon. Mrs. Charles E. B. Russell, one the doctor's greatest admirers and supporters, has left a brief account of four gorillas she encountered while at the hospital in her small book *My Monkey Friends*.[41] She described them as gentle and with table manners not inferior to human children. Similar to so many others, the gorillas didn't survive long or disappeared from the scene after being shipped to Europe.

THE MAN IN A CAGE

Richard Lynch Garner would hunt gorillas in ways no others had or would. Born on February 19, 1848, in Abingdon, a town of about 300 people set within the picturesque Holston River valley of southwestern Virginia, he wound up spending about half his life studying monkeys and apes. What is known of Garner's youth comes mainly from a series of letters he wrote in 1904 addressed to an anonymous friend about growing up.[42] On the surface, nothing in them reveals such a career path in the making. Instead, they show a boy more fascinated with circuses and girls than anything else. Because of being written later in life, the letters reflect an adult re-creating his boyhood. Some of them, though, do ring true about the person Garner became—one skeptical of authority and possessed, almost obsessed, of a willingness to do things differently. For example, he had this to say about education:

> Soon after we were settled in our new home, a young lady, who was my father's cousin, came to live with us and teach a private school in a vacant house that stood in the back yard. This was before the days of public education, in which children are supplied with machine-made education, prepared while they wait, done up in packages of any size, labeled and stamped in four dead languages which the recipient can not read. It was long before the introduction of those modern methods by which active minds of pupils are held in check by impiric ballast while a substitute for brains is furnished to the stupid ones by the farce of written examinations; and thus all minds reduced to a common standard.[43]

And Garner had little compunction about speaking his mind when it came to religion, which he saw as being at a war with the biological sciences. Darwinian evolution undergirded Garner's thinking, and thus he delighted in taking potshots at the religious:

> If there was any taint of carnality in my ambition to be a preacher it was due to the fact that preachers at the time were a privileged class and enjoyed many blessings which the laity didn't even aspire to. The itinerant preacher of my

> boyhood was known as the "circuit rider," which for a long time, confounded with "circus rider"—and on reflection, I think there was more difference in the spelling that there was in the professions, for both of them went around on horseback for the money they could get. Of course, they dressed differently while riding, but aside from that the greatest difference that I could ever see between them was that one of them had no stirrups to his saddle and had to pay his board, and the other had stirrups to his saddle and didn't pay for anything.[44]
>
> The only really bad thing that he [a schoolteacher] did was to leave the Methodist Sunday School and go over to the Episcopalians. Of course, you know they are a kind of neutral church and don't do much harm. . . . They are a free and easy kind of people; they believe in circuses and don't believe that hell is less than a mile away from everywhere, like the Methodists do.[45]

Elsewhere, Garner said that familiarity with domesticated animals created an interest in their vocalizations; he thought they might represent languages or at least the rudiments thereof. The desire to find out remained latent until a cage with monkeys and a large mandrill at the Cincinnati Zoo caught his eye during a visit one day in 1884. Several of the monkeys were keeping a close watch on their much larger co-resident and seemed to signal the troop vocally when he became active. As Garner put it, "From this start I determined to learn the speech of monkeys."[46] He would come to believe in the "psychic unity of all animate nature" and, accordingly, thought that animal languages contained within them the seeds of human speech.[47]

The search for subjects took Garner to zoos in New York City, Philadelphia, and Chicago. He sought out organ grinders and traveling shows and went to private homes where monkeys were kept in order to observe and photograph gestures and record sounds on the new Edison cylinder phonograph. In 1891, Garner first described his results in an article titled "The Simian Tongue." Well aware that attacks from skeptics would be forthcoming, he stated, "I am willing to incur the ridicule of the wise and the sneer of bigots, and assert that 'articulate speech' prevails among the lower primates, and that their speech contains the rudiments which the tongues of mankind could easily develop; and to me it seems quite possible to find proofs to show that such is the origin of human speech."[48] The next year, his *The Speech of Monkeys* appeared on bookstore shelves. Capuchins had been studied most, and he claimed to have identified sounds for things like drink, milk, and food, and he also thought that they could identify colors and "possessed the first principles of mathematics."[49] One monkey seemed to talk about the weather, whereas another complained about cage mates. While he didn't think their language "capable of shading sentences into a narrative or giving any detail in a complaint," they did appear to vocalize "above a mere series of grunts and groans."[50]

During investigations at the Cincinnati Zoo, Garner recorded some chimpanzee sounds, but security measures kept him from making a close study of them. Convinced that "the highest physical types possessed a higher type of speech than those of inferior kinds," he decided to go to Africa in order to examine both chimpanzees and gorillas.[51] As others had, Garner chose Gabon and set on his journey from New York City on July 9, 1892. After a stop in England, he reached Libreville on October 18 and from there went south to the Fernan Vaz region hoping to locate a suitable site for his study. This led him to travel inland along the Ogouwe River for several months before coming back to a place in Fernan Vaz, known for its large populations of chimpanzees and gorillas. Reversing the normal order of things, Garner oversaw the construction of "Fort Gorilla," a steel-wire cage six and half feet square from which he could look out to observe and photograph animals in natural settings as opposed to doing so with them behind bars. Missing, though, was a phonograph. He had planned to get one while in England but couldn't because of patent rights uncertainties.[52] Sounds would thus have to be written down, but how to transpose them into letters eluded him. He thus developed a notation system that proved to be so awkward and indecipherable that afterward he couldn't make sense of his notations.

Garner had the cage painted a "dingy green," and he fashioned a roof from bamboo leaves so as to be less conspicuous to potential visitors. The inside contained a fold-up bed, camp chair, table, supplies, tools, and utensils. For protection, he had a rifle, revolver, and crossbow with prussic acid–tipped arrows. Garner's only companion at Fort Gorilla was a chimpanzee called Moses, which he acquired from a trader on the way up the Ogouwe. People, though, were close by. They included his African servants, porters, and guides; villagers who helped him build the cage; and members of a Catholic mission station.

Garner spent time inside the cage from April 25 to August 6, 1893, but how much time is uncertain. Even less certain is the time devoted to observing chimpanzees and gorillas. Occasionally, he left the cage to track them and made trips to seek out informants, and it's known that he regularly visited the mission station. More certain is the fact that Garner had few close-up opportunities with gorillas. His main chance came when a local man brought him one about six months old. Given the name Othello, he suddenly took ill one day and not too long afterward died. Garner suspected that he had been poisoned and performed an autopsy but failed to identify the cause. Othello's skin went to a museum in Toronto. Later, he secured another young one that also survived for only a short time. Neither gorilla showed the level of sociability displayed by the chimpanzees he had worked with.

In all, Garner reported observing twenty-two gorillas from the cage and one other while out hunting. Given the uncertainties surrounding Du Chaillu's claims, if true, and there's no reason to doubt his veracity, this could represent the largest number seen by any European to date. Most were fleeting glimpses, usually of solitary animals, although a group of ten came briefly into view one day. His longest observation, a female with a baby, lasted about four minutes. Garner said he watched every movement she made and thought about shooting her to get the young one but resisted the temptation. All of his attempts to track gorillas led to failure.

Such limited contact meant that Garner could not really study gorilla sounds in any detail. Only two of them heard, he thought, might qualify as words of some sort. And he really didn't have much to report about gorilla behavior from direct observation. Like those before him, Garner was forced to rely on information provided by Africans and several European hunters. Nonetheless, in several ways, a very different animal emerged in the pages of his 1896 book *Gorillas & Chimpanzees* than most opinion still held. He described gorillas as "shy and timid" and always on the alert for humans in order to avoid contact rather than attack. They didn't roar and scream, and he attributed the thumping sound to beating on a log instead of their chests, something they've been shown occasionally to do. Assertions by some about gorillas throwing sticks and stones at enemies were, he felt, "Mere freak of fancy," and they didn't bend gun barrels after disposing of their victims. He found no evidence for gorillas building houses and he labeled them nomadic, usually traveling in groups numbering ten to twelve. This, too, would prove to be true.

Garner's attempts to describe gorillas in human terms often sound strange to modern ears. Yet kernels of truth can be discerned. For instance, regarding their social life, he stated,

> It is certain that the gorilla is polygamous in habit, and it is probable that he has an incipient idea of government. Within certain limits he has a faint perception of order and justice, if not of right and wrong. I do not ascribe to him the highest attributes of man, or to exalt him above the plane to which his faculties assign him; but there are reasons to justify the belief that he occupies a higher social order than other animals, except the chimpanzee.
>
> In the beginning of his career, in independent life, the gorilla selects a wife with whom he appears to sustain the conjugal relations thereafter, and preserves a certain degree of marital fidelity. From time to time he adopts a new wife, but does not discard the old one; in this manner he gathers around him a numerous family, consisting of his wives and their children. Each mother nurses and cares for her own young, but all of them grow up together as children of one family. There is no doubt that the mother sometimes corrects and sometimes chastises her young, which suggests a vague idea of propriety. The father exercises the function of patriarch in the sense of a ruler, and the natives call him *ikomba njina*, which means gorilla king. To him the others all show a certain amount

> of deference. Whether this is due to fear or respect, however, is not certain, but here is at least the first principle of dignity.
>
> The gorilla family, consisting of this one adult male and a number of females and their young, are themselves a nation. There do not appear to be any social relations between different families, but within the same household there is apparent harmony.[53]

In at least one instance, Garner did go a bit overboard:

> The natives aver that the gorillas from time to time hold palavers or a rude form of court or council in the jungle. On these occasions, it is said the king presides; that he sits alone in the centre, while the others sit in a rough semicircle about him, and talk in an excited manner. Sometimes the whole of them are talking at once, but what it means or alludes to no native undertakes to say, except that it has the nature of a quarrel. To what extent the king gorilla exercises the judicial function is a matter of grave doubt, but there appears to be some real ground for the story.[54]

From what people told him, Garner concluded that gorillas were quite rare. Wrongly, he felt, none lived north of the Ogouwe River, and he added his voice to those who worried that gorillas might soon become extinct, with increasing slaughter by Africans to supply the demands of Europeans being an important contributor.

Garner went back to Gabon many times as much to collect ethnographic information as to study apes. He did manage to send two gorilla skulls off to the Smithsonian, and he observed a dozen or so live captive ones over a span of twenty years. He sent one of them, a tiny female, to the New York Zoological Park. Malnourished, she died on September 23, 1911, a mere two weeks after arrival.[55] Garner described two others in his only other publication about gorillas, a short communication in 1914 sent to the New York Zoological Society. He felt that a female named Dinah showed signs of having a sense of humor. From her pranks, smiles, and laughs, Garner concluded that she was "conscious of being funny."[56] He arranged for her also to be sent to the New York Zoological Park, where she arrived August 21, 1914, only to die less than a year later from malnutrition, including rickets. Dinah apparently lost her zest for life in the new environment and stopped eating. The other, a male named Don, Garner labeled timid and stoic. No record of what happened to him seems to exist.

With regard to gorillas and other primates in captivity, Garner emphasized the need for them to have recreational opportunities and companions; "otherwise," he said, "they become despondent and gloomy." As we shall see, it would be many years before the wisdom of these words would be recognized. As for the man himself, he died on January 22, 1920, with his hometown of Abingdon, Virginia, serving as a final resting place.

No one followed directly in Garner's footsteps, and by the time of his death, the study of language had begun shifting to concerns other than the simian–human connection.[57] But apes and language would eventually come back into the news, albeit in a different way.

NOTES

1. The best source of information concerning Du Chaillu's life is found in Henry H. Bucher Jr., "Canonization by Repetition: Paul Du Chaillu in Historiography," *Revue Française d'Histoire d'Outre-Mer* 66 (1979): 15–32; *Paul Du Chaillu: Gorilla Hunter* (New York: Harper & Brothers, 1930) by Michel Vaucaire is not reliable.
2. Edward Clodd, *Memories* (New York: G. P. Putnam's Sons, 1916), 71.
3. See Barbara Lynn, *The Heyday of Natural History, 1820–1870* (Garden City, NY: Doubleday & Company, 1980).
4. Paul D. Du Chaillu, "Descriptions of the Habits and Distribution of the Gorilla and Other Anthropoid Apes," *Proceedings of the Boston Society of Natural History* 7 (1860): 296–304, 358–67.
5. Heinrich Barth, "Analyse der Reisebeschreibung du Chaillu's *Explorations and Adventures in Equatorial Africa*, und genanere Betrachtung des in derselben enthaltenen geographischen Materials," *Zeitschrift fur allgemeine Erdkunde* 10 (1861): 430–67.
6. For controversies surrounding Du Chaillu's travels and discoveries, see Joel Mandelstam, "Du Chaillu's Stuffed Gorillas and the Savants of the British Museum," *Notes and Records of the Royal Society of London* 48 (1994): 227–45, and Stuart McCook, "It May Be Truth, but It Is Not Evidence: Paul du Chaillu and Legitimation of Evidence in the Field Sciences," in "Science in the Field," ed. Henrika Kuklick and Robert E. Kohler, *Osiris* 11 (1996): 177–97.
7. R. B. Walker, "M. Du Chaillu and His Book," *The Athenaeum* 21 (September 1861): 374; McCook, "It May Be Truth, but It Is Not Evidence," 189.
8. Edward Bartlett, ed., *Wild Animals in Captivity* (London: Chapman and Hall, 1889), 5.
9. Ibid., 131.
10. K. David Patterson, "Paul B. Du Chaillu and the Exploration of Gabon 1855–1865," *International Journal of African Historical Studies* 7 (1974): 647–67.
11. Paul Du Chaillu, *Explorations and Adventures in Equatorial Africa* (London: J. Murray, 1861), 70–71.
12. Ibid., 276.
13. Ibid., 297.
14. Ibid., 244.
15. Ibid., 347–48.
16. Cited in D. B. Willoughby, *All about Gorilla*s (Cranbury, NJ: A. S. Barnes and Co., 1979), 40.
17. W. Winwood Reade, *Savage Africa* (New York: Harper & Brothers, Publisher, 1864), 179–80.

18. W. Winwood Reade, "The Habits of the Gorilla," *American Naturalist* 1 (1867): 177–80.

19. Thomas H. Huxley, *Man's Place in Nature and Other Anthropological Essays* (New York: D. Appleton and Company, 1909), 72.

20. Paul Du Chaillu, *A Journey to Ashango-Land* (London: J. Murray, 1867), 54.

21. Adolphus W. Greeley, *Men of Achievement, Explorers and Travelers* (New York: Charles Scribner's Sons, 1893).

22. Richard F. Burton, *Two Trips to Gorilla Land and the Cataracts of the Congo*, vol. 1 (London: S. Low, Marston, Low, and Searle, 1876).

23. W. Garden Blaikie, *The Personal Life of David Livingstone* (New York: Laymen's Missionary Movement, 1890), 424.

24. Robert Hamil Nassau, *In an Elephant Corral* (New York: Neal Publishing Company, 1912).

25. P. T. Barnum, *Struggles and Triumphs: or, Forty Years' Recollections* (New York: American News Company, 1871), 692–96.

26. John Fortune Nott, *Wild Animals Photographed and Described* (London: Sampson, Low, Marston, Searle, & Rivington, 1886), 534–38.

27. Henry Scherren, *The Zoological Society of London: A Sketch of Its Foundation and Development* (London: Cassell and Company Limited, 1905), 202.

28. Robert M. Yerkes and Ada W. Yerkes, *The Great Apes: A Study of Anthropoid Life* (New Haven, CT: Yale University Press, 1929), 384.

29. Sir Arthur Keith, "An Introduction to the Study of Anthropoid Apes. 1. The Gorilla," *Natural Science* 9 (1896): 26–37.

30. Cited in Nott, *Wild Animals Photographed and Described*, 533.

31. Quoted in Michael Nichols and George B. Schaller, *Gorilla Struggle for Survival in the Virungas* (New York: Aperture Foundation, 1989), 9–10.

32. Willoughby, *All about Gorillas*, 152.

33. For examples of carved face masks, see Leon Siroto and Kathleen Berrin. *East of the Atlantic West of the Congo: Art from Equatorial Africa* (Seattle: University of Washington Press, 1995), 30. Unfortunately, I was unable to gain permission to reproduce the image.

34. For overviews, see Horst W. Janson, *Apes and Ape Lore in the Middle Ages and the Renaissance* (London: Warburg Institute and University of London, 1952), 261–86, and Ramona Morris and Desmond Morris, *Men and Apes* (New York: McGraw-Hill Book Co., 1966), 54–82.

35. Cited in Alfred Edmund Brehm, *Brehm's Life of Animals, Mammalia,* vol. 1, 3rd ed. (Chicago: A. N. Marquis & Company, 1895), 14.

36. Ibid.

37. Ibid., 3.

38. Fred G. Merfield and Harry Miller, *Gorillas Were My Neighbors* (London: Longman, Green, 1956), 73–76.

39. Ibid., 100–101.

40. Fred G. Merfield and Harry Miller, *Gorilla Hunter* (New York: Farrar, Straus and Cudahy, 1956), 223.

41. Mrs. Charles E. B. Russell, *My Monkey Friends*, 2nd ed. (London: Adam & Charles Black, 1948), 56–60.

42. Harry E. Garner, *Autobiography of a Boy, from the Letters of Richard Lynch Garner* (Washington, DC: Huff Duplicating Company, 1930).

43. Ibid., 12.

44. Ibid., 72.

45. Ibid., 82–83.

46. Richard L. Garner, *Apes and Monkeys: Their Life and Language* (Boston: Ginn & Company, 1900), 15.

47. Ibid., iv.

48. Richard L. Garner, "The Simian Tongue [I]," *New Review* 4 (1891): 555–62.

49. Richard L. Garner, *The Speech of Monkeys* (New York: Charles L. Webster and Company, 1892), 28.

50. Ibid., 35–36.

51. Richard L. Garner, *Gorillas and Chimpanzees* (London: Osgood McIlvaine and Co., 1896), 14.

52. For more details on Garner's methods, claims, and reactions to him, see chapters 3 and 4 in Gregory Radick, *The Simian Tongue: The Long Debate about Animal Language* (Chicago: University of Chicago Press, 2007).

53. Garner, *Gorillas and Chimpanzees*, 214–15.

54. Ibid., 216.

55. Letter dated April 17, 1928, from W. Reid Blair, director of the New York Zoological Park, in Yerkes and Yerkes, *The Great Apes*, 383–84.

56. Richard L. Garner, "Gorillas in Their Own Jungle," *Zoological Society Bulletin* 17 (1914): 1102–4.

57. An excellent overview of the changing tenor of the apes and language debate can be found in Radick, *The Simian Tongue*.

Chapter Four

In Search of Mountain Gorillas

As the nineteenth century drew to a close, the Mountain gorilla remained unknown to the outside world. The many travelers and explorers passing through this part of Africa skirted their homelands, which are nestled in an area some forty miles long and no more than twelve miles wide defined by the Virunga Volcanoes. Six are inactive—Mikeno (14,557 ft., 4,437 m), Karasimbi (14,787 ft., 4,507 m), Visoke (12,175 ft., 3,711 m), Sabinyo (11,959 ft., 3,645 m), Gahinga (11,400 ft., 3,474 m), and Muhavura (13,540 ft., 4,127 m)—while two near Lake Kivu, Nyiragongo (10,385 ft., 3,470 m) and Nyamuragira (10,033 ft., 3,058 m), continue to flare up from time to time. An eruption by Nyiragongo on January 10, 1977, buried villages and killed an unknown number of people. Another, on January 17, 2002, destroyed parts of the city of Goma, which had to be evacuated. Some of the travelers, though, did hear stories of large, dangerous creatures inhabiting the forested slopes. In December 1861, King Rumanika of Karagwe told John Hanning Speke and James Grant during their search for the source of the Nile River about "monsters" living in Rwanda "who could not converse with men, and never showed themselves unless they saw women pass by; then, in voluptuous excitement they squeezed them to death."[1] During his 1873–1875 journey across Central Africa, Verney Lovett Cameron claimed he saw gorillas along the northwest shore of Lake Tanganyika but that they quickly disappeared before he could get a shot off at them.[2] Given the location, it was likely a troop of chimpanzees that caught his eye.

At this point in time, it would seem that only Batwa hunter-gatherers knew of Mountain gorillas from firsthand experience, having encountered them while searching for honey and small game, such as hyrax and duiker. That they were the ones who concocted the tall tales of gorilla ferocity is suggested by a response that George Schaller received in 1959 from a Batwa tracker after telling him about a Bahutu farmer who claimed that gorillas "grab flying spears and throw them back at the attacker." The tracker replied, "This is a fable. We tell such tales to the Hutu and they believe them. The gorillas fear man. They bark and roar and run away."[3] The Batwa are at home in the forest. It's their source of sustenance, and they wanted to keep it that way. To the Bahutu and Batutsi, the forest signifies danger and thus something to be cut down and turned into fields and pasture whenever possible.[4]

The first non-African to see direct evidence of the Mountain gorilla appears to have been Colonel Ewart S. Grogan while on a Cape-to-Cairo sojourn in 1898. During a hunt for elephants north of Lake Kivu, he reported coming across "the skeleton of a giant ape, larger than anything I have ever seen in the anthropoids." According to local testimony, many such animals existed in the mountains "and were a great source of annoyance to the villages, being in the habit of carrying off stray women."[5] Because Grogan didn't collect the skeleton, confirmation of the Mountain gorilla fell to Captain Oscar von Beringe. In October 1902, he made the second of two trips to Virunga country while in the service of German East Africa. According to him, the task this time was designed

> to keep the tribal chiefs in sympathy with the German government and to further firm up, right then and there, the good relations with the Sultan of Ruanda cultivated by my predecessors. The march at the volcanoes served not only to safeguard the Congolese Border Commission that is now at work here, but, at the same time, to strengthen the power of and regard for the German administrative authority in these territories.

While encamped one day, the party saw

> a herd of large, black apes attempting to climb the highest peak of the volcano. We succeeded in bagging two large animals from among these apes; with a great rumbling they had plunged into a crater gorge opening out to the northeast. After five hours of heavy labor, we managed to pull one animal up by rope. It was a large, masculine, human-like ape about 1–1.5 m. tall and weighing over 200 lbs. Hairless chest, hands and feet of immense size. Unfortunately, it was impossible for me to determine the ape's species. For a chimpanzee it would no doubt have been of a size still unknown at that time, and the presence of gorillas [from here] to the lakes has not been established up to now.[6]

Von Beringe sent the specimen to Germany, where zoologist and taxonomist Paul Matchie identified it as a subspecies of gorilla he labeled *Gorilla gorilla beringei*.[7] His perception of distinctiveness has been validated many times over. That same year British Lieutenant Colonel R. H. Elliot forwarded two skulls obtained from a German engineer to the Museum of the Royal College of Surgeons in London. They, too, turned out to be those of Mountain gorillas.

Similar to what happened to the Western gorilla, discovery set off a round of trophy hunting and collecting. The South African Philip J. Pretorius killed one in 1904 just for the thrill of it, something he often did on hunts for many other animals:

> I looked out on the clearing in the bush, roughly fifty yards in diameter, where the bamboos had been trampled flat by the herd of elephants. In the centre of the clearing sat a huge hairy object lighted up fitfully by the sun, which was now beginning to penetrate the mists. A huge head turned towards our direction, and one of the most diabolical faces I have ever seen held me spellbound; a face with no forehead, huge, overhanging eyebrows, two small, close-set eyes, glinting redly, a flattened nose with large nostrils, and, finally, a huge slit of a mouth, bulging with teeth. The unprepossessing beast was chewing young shoots of bamboos torn down by the elephants. I was absolutely fascinated by the ogre, and was only brought back to reality when the wind wafted to my senses a most pungent smell which caused both my nostrils and throat to smart. Slowly, I raised my rifle, aiming at the base of the skull, and as I pressed the trigger the beast collapsed without a murmur.[8]

Pretorius justified this less than sporting kill by noting that at the time "both Europeans and natives regarded the gorilla as a most ferocious and dangerous animal, liable to attack any hunter on sight."[9]

In 1913, Elias Arrhenius went looking for specimens to ship to Stockholm's Riksmuseum. His party managed to shoot seven in and around Mount Mikeno, although the outbreak of World War I delayed their arrival at the museum until 1916. In the same year, Commandant Powells of the Belgian army secured the first specimen from the forests to the west of Lake Kivu. It went on display at the Museum for Central Africa located in Tervuren, Belgium, billed as a Mountain gorilla.[10] That area, however, is Eastern, not Mountain, gorilla territory. Others would soon make the same mistake.

Naturalist T. Alexander Barns ventured into the Congo portion of Virunga country in 1919 during a sightseeing and collection trip that started in the Katanga region of the Congo and ended with a journey down the Congo River. Along the way, he shot an old male that stood over five feet in height and weighed approximately 450 pounds but passed on another opportunity to make a kill.[11] The carefully preserved remains made it safely back to England, where they were mounted for display, eventually winding up at Lionel Walter's Rothschild Museum in 1923, courtesy of the Rowland Ward Trustees. Ward was a famous taxidermist specializing in big game and through his business and publishing had amassed a sizable fortune by the time of his death in 1912.

Barns's objectives centered more on photographing and observing animals than killing them. As for gorillas, he concluded that they were not, as many people believed at the time, arboreal. Indeed, he saw nary a one in trees. As noted previously, they do, however, climb them. And from what he could tell, Mountain gorillas appeared to have no enemies other than humans, and even they posed little threat because contacts were few due to the dense forested environment in which the apes lived.

Barns returned to the Virungas in 1921 in order to obtain specimens of Mountain gorillas and a rare chimpanzee found in the Kivu area for the British Museum, which, of course, meant killing them. He succeeded with both objectives, the gorillas being a male, female, and juvenile, to provide a family display. In this instance, Barns had an opportunity to observe gorillas more thoroughly, reporting on foods eaten and nest making. Contrary to local opinion, he called the gorilla a "great bluffer, and if he can't frighten you away by his uncanny screaming roars or by the beating of his great chest, he leaves it at that—he is certainly not looking for trouble."[12] Barns also felt that the value of all surviving great apes rested primarily with what could be learned scientifically from them. As such, he felt that "they should be rigorously protected."[13]

Between Barns's two trips, the Virunga gorillas had to face an expedition led by Sweden's Prince William. A veteran hunter, the prince, like so many others at the time, saw Africa as an escape from the confines of civilization, a place to discover the true meaning of being human. According to him, "He who has got the wilderness into his blood, has lived its life, fought its fights, and enjoyed its joys is lost. For he longs to return with a longing that nothing can appease and which grows with the distance."[14] Gorillas provided another opportunity for the prince to play out his fantasies cum desires.

The prince, too, would travel under the guise of collecting for the Riksmuseum, and thus he took along Count Nils Gyldenstolpe to make illustrations and Oskar Olsson to take photographs. Veteran big-game hunter Kenneth Carr served as safari guide. They had permission from Belgian authorities to shoot fourteen gorillas and without hesitation filled their quota. The story told by William in *Among Pygmies and Gorillas* focused on the thrills of the hunt, in this instance not so much the danger posed by gorillas. In fact, he found them far less dangerous than lions, elephants, rhinos, and Cape buffalos. Rather, the thrill involved tracking and locating specimens. For the most part, Gyldenstolpe concurred, calling the hunt "one of the most difficult and fatiguing sports in existence, but when you succeed the gorilla is a worthy object."[15]

As for findings, the expedition provided the first detailed descriptions of gorilla sleeping nests and confirmed that the Mountain version had longer

Plate 4.1. A gorilla bust in bronze by Herbert Ward. *Source:* Department of Anthropology, Smithsonian Institution, catalog E323720.

hair than its Western cousin. The members also heard the same stories about gorillas doing such things as building huts with roofs, making fires, stealing women, and eating children. The widespread nature of such stories is yet another indication of how highly interconnected African societies were before Europeans arrived on the scene.

Art at the time seems to have had trouble freeing itself from Du Chaillu. An example is a bust produced by Herbert Ward. He had spent many years in the Congo during the latter part of the nineteenth century and was a member of the ill-fated "rear guard" of Henry Morton Stanley's infamous Emin Pasha Expedition of 1887–1889. Ward later turned to sculpting and created a number of highly regarded pieces based on his Congo experiences.[16] The only one of an animal is a natural-size bronze of a gorilla head with mouth open wide to reveal long, protruding canines. The visage, too, is one suggesting a beast. How he came to this particular representation is unclear. Did he ever see a live gorilla while in the Congo? Use a specific stuffed one as a model? Create it based on sources he had consulted? We just don't know. In 1921, Ward's wife Sarita (he died in 1919) donated the piece to the Smithsonian in Washington, D.C., with the accession form stating that it "well expresses the bestial ferocity of this strange creature."

Deeper understanding of the Mountain gorilla came from an unlikely source, Carl Akeley. Born on May 19, 1864, and christened Clarence Ethan Akeley, he spent his childhood on the family farm in Clarendon, New York.[17] Pretty much a loner and not very interested in farming, Carl frequently went off on nature hikes by himself. Animals provided a particular fascination, and a trip to nearby Rochester introduced him to taxidermy. He bought a book to learn its rudiments and began practicing on birds. Carl also took painting lessons in order to create natural settings for their display. At age sixteen, he went to school in Brockport and while there got a chance to see painter and taxidermist David Bruce at work. Bruce urged him to further his skills at Ward's Natural Science Establishment in Rochester, at that time the country's most famous taxidermy workshop and responsible for many of the animals on display at natural history museums. Akeley's tenure there in 1883 didn't go well. Fussy about details, he made enemies by criticizing the work of others and frequently talked back to superiors, even to the eminent owner Professor Henry Augustus Ward. Akeley chafed at the prevailing practice of simply mounting specimens for customers, feeling that taxidermy should be more naturalistic. This meant paying careful attention to anatomy and displaying animals in replicas of their natural settings. After just a few months at Ward's, Akeley received his walking papers for acts of insubordination, including sleeping on the job.

He next worked for John Wallace, a commercial taxidermist in Brooklyn. The stint deepened Akeley's frustrations with the field. It struck him as nothing more than a trade, and the time spent in the basement workshop he described as a "dreary six months," working mostly at getting birds ready for display on women's hats.[18] These were all the rage in the late nineteenth century. When an opportunity to return to Ward's came up, Akeley jumped at it. Not too long afterward, P. T. Barnum's massive and hugely famous African elephant Jumbo died from being hit by a freight train in St. Thomas, Ontario.[19] Barnum immediately contacted Ward about doing a reconstruction for circus display. Unable to go at the time, Ward sent Akeley in his stead. After carefully stripping and preserving the skin from the corpse and packing up the skeleton, Akeley returned to his workplace in Rochester and several months later produced a likeness of Jumbo that amazed everyone. In this guise, the great elephant accompanied the circus for many more years. Today, his remains are at the American Museum of Natural History.

Akeley made a friend at Ward's, William Morton Wheeler from Milwaukee. Tired of taxidermy, Wheeler returned home to become a teacher and work with the city's cultural museum. The two men stayed in touch, and in November 1886, Akeley headed for Milwaukee to take a position at the museum and do taxidermy on the side. Contracts came regularly, especially after Wheeler became the museum custodian in September 1888. Two years later, Wheeler left to take a scholarship at Clark University in Worcester, Massachusetts. Akeley stayed on in Milwaukee until 1892, when he went to DeKalb, Illinois, to set up his own studio. At year's end, he received his biggest commission yet—to create three horses for a Native American display at the World's Columbian Expedition, set to open in Chicago on May 1, 1893. His work won unanimous praise for its lifelike qualities and led to an offer to join the newly established Field Museum in Chicago, then headed by the renowned natural historian Daniel G. Elliot. Akeley accepted and went about creating a new studio along with his wife-to-be, Delia (Mickey) Reiss. The two had met in Milwaukee when she was the teenage bride of a barber. A romance quickly blossomed, and Delia left her husband to be with Akeley.

Shortly after being put on the payroll, Elliot asked Akeley if he would like to go with him on a collecting expedition to Africa. Ignoring warnings of danger because of unsettled conditions, they chose British Somaliland as the point of departure. During a sojourn that lasted from April 21 to October 1, 1896, the expedition collected a bounty of around 200 mammal skins, including oryx and kudu; an even greater number of birds, reptiles, and fish; and 300 photographs. Elephants, buffalo, and rhinos eluded them, however. Despite almost dying twice, once from thirst and another time at the jaws of a leopard, which he somehow managed to kill with his bare hands, Africa had gotten under Akeley's skin to the extent that it would soon become his lifelong passion.

A return, though, would have to wait, as other work beckoned. One assignment involved a trip to the Pacific Northwest to secure specimens like elk and wapati. Back in Chicago, he and Delia spent their days working on these and the African ones. Spare moments were devoted to creating an exhibit of deer in natural settings during spring, summer, fall, and winter, a project they had begun before the trip to Somaliland. Called the Four Seasons, they completed it in 1902 for display at the Field Museum. In December of that year Akeley and Delia officially became man and wife.

On August 13, 1905, Akeley left Chicago, bound for East Africa on a second Field Museum expedition, this time with Delia at his side. They chose Nairobi, a railway camp blossoming into a city, as the base of operations for an expedition aimed at large animals, elephants in particular. In addition to shooting specimens to mount, Akeley put in many days studying them so that his recreations would appear lifelike. Eventually, they killed two bull elephants, she bringing down a huge one in the forests of Mount Kenya. An array of animals joined the elephants, and, when put together with other specimens needed for museum dioramas, the expedition returned to the United States in early 1907 with eighty-four crates of remains weighing some seventeen tons. Later, Akeley had misgivings about what had taken place, noting that shooting animals, save lions and elephants, which he figured "had sufficient chance in the game," for expositions made him feel "a great deal like a murderer."[20]

Bigger things now awaited Akeley, as he accepted an offer to join the American Museum of Natural History in New York City. Its president, the world-renowned vertebrate paleontologist Henry Fairfield Osborn, wanted an elephant exhibit to add to the museum's spectacular dinosaur collection, and Akeley agreed to lead an expedition to secure the necessary specimens. Delia would again accompany him, and in September 1909 they were back in Nairobi. In addition to elephants, Akeley wanted to do some filming of a lion hunt by Nandi warriors.

The trip produced a meeting with Theodore Roosevelt, who had been impressed with the Four Seasons exhibit during a visit to the Field Museum. Akeley was away at the time, but the two men met in late 1906 when the president requested his presence at a White House dinner. The idea of an African hunting trip together came up for discussion. Roosevelt wanted to go on one when his term in office ended in 1909, both for reasons of sport and to collect specimens for the Smithsonian in Washington, D.C. By the time Roosevelt could get around to making arrangements, Akeley had committed to the American Museum of Natural History's expedition.

Roosevelt left first, and, as it turned out, the two men's paths crossed near Mount Elgon. They had a chance to go elephant hunting together and talk intimately. This proved to be the highlight of the trip, as week after week went by in a futile search for the big bull that Akeley sought. In addition, problems with the porters surfaced regularly, sometimes leading Akeley to using a whip as punishment, an old tactic dating to the days of the slave caravans. His health took a beating from numerous bouts of malaria, dysentery, a nearly fatal attack of the blackwater fever, and a painful case of jiggers. Then, on July 24, 1910, a bull elephant mauled him while hunting the slopes of Mount Kenya. Fortunate not to have been gored or trampled to death, he lay unconscious for hours, the men, in fact, thinking him dead. Delia was at their base camp, and when news arrived of what had happened, she browbeat some of the men there into going with her to find Akeley. With the help of a Scottish medical missionary who came along, they got him back to camp, and after three months of devoted wifely care, he managed to get back on his feet and resume the search for that elusive big bull elephant. At last, Akeley came across what he had been after, and with Delia's help they made the kill. But the fabric of their relationship was beginning to unravel. She had acquired a vervet monkey named "J.T., Jr." who eventually replaced Akeley at the center of her life.[21]

While recuperating from the elephant mauling, Akeley conceived of a "Great Hall" at the American Museum of Natural History to exhibit African wildlife in their natural settings. More than taxidermy, though, would be required to achieve what he wanted. This caused him to reanimate an interest in sculpture, and when back home he set about creating in bronze. His first production, called *The Wounded Comrade*, showing a wounded bull elephant being assisted by two others, earned him a membership in the American Sculpture Society. Over the years, he produced six other highly acclaimed bronzes.

The outbreak of World War I put a return to Africa on hold, so Akeley turned his creative energies elsewhere. He had managed to film the Nandi lion hunt but was dissatisfied with the outcome. This led him to form a company to produce a camera that would take better images of action scenes. He succeeded, and the Akeley Motion Picture Camera quickly became the standard for adventure photography. Hollywood filmmakers adopted it, as did the army for use in the war. Pathé News and Fox Movietone News also joined the list of enthusiasts.

As Akeley told it, the next opportunity to go to Africa arose during a discussion with his friend Herbert E. Bradley one night in 1920. Bradley wanted to take his family, including five-year-old daughter Alice, to Africa for a year. Akeley assured him that Central Africa would be perfectly safe, and he assembled a party consisting of the three Bradleys, another friend of Akeley's, Martha Miller, and Priscilla Hall, who would care for Alice when the others went hunting. They were destined for gorilla country in Rwanda. Although he had never seen one, Akeley had developed a "creed" about the Mountain gorilla that he related to the others going with him:

> I believe that the gorilla is normally a perfectly amiable and decent creature. I believe that if he attacks man it is because he is being attacked or thinks that he is being attacked. I believe that he will fight in self-defense and probably in defense of his family; that he will keep away from a fight until he is frightened or driven into it. I believe that, although the old male advances when a hunter is approaching a family of gorillas, he will not close in, if the man involved has the courage to stand firm. In other words, his advance will turn out to be what is usually called a bluff.[22]

He wished to accomplish three things during the expedition—shoot specimens for the Great Hall, study gorillas, and film them so as to have a record to consult when creating the diorama. The successes of Prince William and professional hunter C. D. Foster in bagging Mountain gorillas gave him confidence that these goals could be achieved.

In spite of Akeley's "creed," general opinion still considered gorillas dangerous. Mary Bradley said that friends tried to scare her with one man's supposed fate by saying, "The poor brave fellow who had gone off alone was lying on the ground in a pool of his own blood, his entrails torn out, his gun beside him, bitten in two by the gorilla's teeth."[23] Such words, however, failed to make a dent in her determination to go as planned.

The party departed from New York harbor on July 30, 1921, for England and thence Cape Town. From there, they took trains to Bukama in the Katanga region of the Belgian Congo and then sailed down the Lualaba River in a steel barge towed behind a river steamer. At Kabalo, a train carried them to Albertville (now Kalemie) on the western shore of Lake Tanganyika for a boat ride to Usumbura. A seven-day hike led to the southern end of Lake Kivu to pick up another boat. During a stop at a White Fathers mission station, they heard about gorillas having recently raided a banana plantation. Chased out by villagers, the male reportedly attacked and killed one of the men, after which they speared it to death. Killing a gorilla was an unusual event, as legend held that the act would bring childless daughters and the loss of a son's son.[24] At Kissenyi along the north end of the lake, they met with Mrs. T. Alexander Barns. Her husband was off on the gorilla hunt described earlier.

Akeley left the others at Kissenyi for the Virungas on the last day of October. Mrs. Bradley had fallen ill, and he didn't want to go into gorilla country with a large party. After several days of rough going, he caught sight of a single male and set off after him. Slowly moving closer, Akeley fired a lethal shot, and the gorilla tumbled down a steep slope. Presumed lost, the gorilla instead had become wedged in a tree. Because of its position, skinning had to be done on the spot.

After a day spent preparing the specimen, Akeley resumed his hunt despite being exhausted and down some twenty pounds in weight. He picked up a trail but missed several opportunities. Then an old female came into clear view. Akeley fired and fell to the ground from the rifle's recoil. She went over him, and then, as he started to rise, "there seemed to be an avalanche of gorillas" speeding by.[25] None of the gorillas attacked, convincing him that they were not ferocious. As he later wrote, the gorilla is a "good-tempered beast, who expresses himself by a bark rather than a terrifying roar."[26] Besides killing the female, he wounded a "big ball of black fur" and found it to be a young male that the guides dispatched with spears. Akeley expressed some remorse over the latest kill. "There was a heartbreaking expression of piteous pleading on his face. He would have come to my arms for comfort."[27]

With two more specimens to prepare, it took several days before the next hunt could begin. Akeley first photographed a female with her young one and then followed a band, killing what looked to be a male, although it turned out to be a female. His regret this time was "considerably lessened" by her large size.[28] With four specimens that included old and young males and two females, he had enough for his Great Hall exhibit. Akeley then made death masks of each and plaster casts of their feet and hands. He also had shot several hundred feet of motion picture film, the first ever of free-living gorillas, to make him happy.

The rest of the party now required having their special gorilla adventures. Akeley chose to set up camp in the saddle between mounts Mikeno and Karisimbi, known both for their beauty and their numbers of gorillas. Only Priscilla and Alice stayed behind at Kissenyi. During an episode of filming, a huge silverback appeared. Herbert Bradley fired and hit him, but the gorilla disappeared into the forest. They eventually tracked him down, and Bradley's second shot proved deadly. Known thereafter as the "Lone Male of Karisimbi," according to Mary Bradley, "There had been no sound from him, no bark or roar. He had shown no instinct to fight, nothing but the rush of a wounded animal to escape."[29] Akeley thus had a fifth gorilla to prepare. As in the other instances, the Africans wouldn't eat the meat. That didn't stop the curious Americans, however. As noted by Mary Bradley, "We ourselves had cooked and eaten a little, just for the sake of doing it, and found it perfectly good meat, firm and sweet, but I couldn't get over the family feeling of sampling grand-uncle Africanus!"[30]

While returning to the mission station, Akeley and Herbert Bradley came across a large troop with maybe four males, a dozen females, and their young. Struck by the peacefulness of the scene, Akeley concluded that there was "really no necessity for killing another animal."[31] The others in the safari going via a different route also managed to have an encounter, seeing maybe six gorillas. Mary Bradley described the male as looking directly at them:

> It was an uncannily human face that he turned up to us, but there was none of Du Chaillu's demon horror about it. I got an impression of a wary interest that did not intend to tolerate any intrusion, but there was not a flash of menace—nothing that the most prudish person could possibly call hellish. He simply conveyed the idea that he had been disturbed by a distinct outsider, and started deliberately away, shambling along through his ancestral meadows towards an arch in the tree leading into glades ahead.[32]

They sent a message to Akeley to come and film the group, but it disappeared into the forest, never to be found.

Another priority now captured Akeley's attention—Mountain gorillas needed official protection in special reserves in order to avoid extinction. He thought only fifty, maybe 100, individuals occupied Virunga country and that the three mountains of Mikeno, Karisimbi, and Visoke would provide an excellent sanctuary "where the gorillas under the protection of man may grow more and more accustomed to human beings and where through a series of years they may be observed and studied."[33] Such a place would also serve as balm for the wounds he carried for having to kill animals with the hope of saving them.

The safari moved on for the others to do more hunting. The Bradleys and Martha Miller each got their lion. On New Year's Day 1922, they left the Congo, going via Uganda to Nairobi and then Mombasa to sail for home.

The gorilla diorama had by now taken shape in Akeley's mind. The setting would be modeled on the one surrounding the Lone Male of Karisimbi, and the grouping would display the animals in a way far different from any exhibit so far presented to the public. He considered the others on display ugly and too dependent on Savage and Wyman's earlier description of physical form:

> Had the American Museum undertaken to prepare a gorilla group five years ago, using skins which could be purchased in the open market, and planning the group as carefully as possible in accordance with the accumulated data of the past seventy years, I have an idea that the group would have had a much greater appeal to a public thirsting for excitement and sensation than the group which will result from the knowledge recently acquired. Such an imaginary group would of necessity have shown the gorilla as a ferocious creature in a setting

> of gloomy forest or mysterious jungle. There would have been one specimen in a tree, another walking erect with a staff or club in one hand, and perhaps a third beating its breast with its fists and opening its cavernous mouth as though roaring with rage. A house or nest, ingeniously constructed somewhere between earth and sky, would have been required to make the picture complete. Taking the records literally, there would have been justification for depicting an old male in the act of crushing with his teeth the barrel of a hunter's gun.[34]

Plans for a sanctuary continued to form in Akeley's mind. He began writing to Belgian authorities about the idea, and on January 18, 1923, he presented a series of recommendations to John C. Merriam of the Carnegie Institute in Washington, D.C. Merriam forwarded them to Belgium's ambassador to the United States, who agreed to work on behalf of a sanctuary. The idea resonated with King Albert. He had been thinking about establishing animal reserves in the Congo, and on March 2, 1925, the king issued a royal decree creating the Parc National Albert around mounts Mikeno, Visoke, and Karisimbi. Within its borders, private hunting expeditions and those intended to supply specimens for natural history museums and scientific research would be prohibited. Gorillas would join other animals on the list of royal game for which

> No person who is not holder of a scientific license or is acting in legitimate self-defense may kill, capture, hunt or pursue, intentionally disturb or drive away by any means whatsoever or for any reason whatsoever including the purpose of photographing or filming any of the animals. . . . Persons acting in legitimate self-defense must notify authorities immediately of any such action and surrender the remains of any animal.[35]

In the meantime, the estrangement between Akeley and his wife had deepened. Without notice, she left for Europe in 1918 to work in canteens serving the American Expeditionary Force and upon returning home never went back to him. Both later filed for divorce on grounds of desertion, with the marriage officially ending on March 22, 1923, cruelty having been added by her to the proceedings.[36] She remarried in 1939, but history remembers her as Delia or Mickey Akeley. Death came on May 22, 1970, at the ripe old age of ninety-five.

On October 18, 1924, Akeley married Mary L. Jobe, whom he had met in 1921 (or perhaps earlier—it's unclear). In addition to her, two other things kept him occupied: work on the Great Hall and planning a return to the Virungas in order to do a more thorough gorilla study. An opportunity arose to achieve both ends when George Eastman expressed a desire to go on a hunting safari in Africa. Akeley agreed to accompany him for a fee to help fund completion of the Great Hall and insisted that no gorillas could be killed. Daniel E. Pomeroy, trustee of the American Museum of Natural History, and Colonel Daniel B. Wentz, a prominent Philadelphian, also provided support for the project.

The Akeleys left New York City for London on January 30, 1926, taking along an enormous load of supplies, including vehicles specially designed in Canada. William R. Leigh and Arthur A. Jansson accompanied them to work on the diorama. In Brussels, they obtained permission to conduct an environmental survey of the Parc and its environs and take a census of gorillas, tasks to be done by the Belgian zoologist Jean Marie Derscheid, who would join the expedition in Nairobi. First, though, Akeley decided to visit the Lukenia Hills and Northern Frontier District of Kenya to hunt additional specimens needed for the Great Hall. The thinness of wildlife on the ground compared to his earlier visits shocked him. All of the hunting safaris crisscrossing the countryside had clearly taken their tolls. Next came a trip to the Serengeti in Tanganyika for lions, reputedly the last place where they existed in substantial numbers. During five months in Kenya, none had been seen. Akeley also wanted to film another lion hunt, this time by the Lumbwa or, more accurately, Kipsikis. Lumbwa is what the Maasai insultingly call herding peoples who have taken up farming. Once there, he joined Eastman to capture what amounted to a staged event. For the most part, the two men worked independently, as, in terms of personality, they proved to be water and oil.

Past injuries, age, and new illnesses had by now worn Akeley down, and a wild-dog hunt on the Serengeti safari proved to be, Mary Jobe said, "his *last active day in the field in Africa*."[37] Despite a high fever, he didn't want to give up, but she felt they must, and when the fever subsided, she decided they had to make the 300-mile-long journey to Nairobi to see what ailed him. Exams yielded nothing specific, but three weeks in a nursing facility and a couple more at their home base restored his strength enough to allow the journey to gorilla country to get under way on October 14, 1926. Eastman, Pomeroy, and Wentz had already set off for home, having dropped any association with the Akeley party after filming the lion hunt.

Roads of varying quality existed as far as Kabale near the Rwanda border. From here, the rest of the journey proceeded on foot. Facing sun, rain, mud, and up-and-down trails, Akeley weakened. Still, at his insistence, the trek continued. By the time they set up camp at Kabara in the saddle between Mikeno and Karisimbi, he was in excruciating pain and began hemorrhaging. Peace came on November 17. Five days later, in a coffin made of mahogany lined with metal from used containers, he was laid to rest in an eight-foot-deep grave dug into gravel and lava rock. A flat cement marker stating "CARL AKELEY, November 17, 1926" served as a headstone. Mary Jobe and the rest of the party decided to stay on to complete his work. She concentrated on taking photos while Leigh finished the painting for Akeley's gorilla diorama. Derscheid stayed longer in order to complete his environmental survey and census. Overall, he estimated that the Virungas, including the portion in Uganda, housed between 450 and 650 gorillas, the vast majority being within the Parc boundaries.[38]

On December 19, Mary Jobe began her journey out. On the way home, she visited King Albert and in September 1928 returned to Belgium and helped Derscheid draw up a memorandum about an expanded Parc to include the whole volcanic chain of 500,000 acres. With vigorous support from Prince Albert de Ligne, it received the royal stamp of approval on July 9, 1929. Within the new boundaries, nothing could be taken, killed, or destroyed. Those parts already occupied by villagers or under concession could continue as such, but any killing of gorillas was absolutely forbidden. In selected adjacent territories, customary hunting rights were permitted but only with traditional weapons, not guns. In 1929, the British also set aside Uganda's portion of the Virungas as protected territory.

Between Carl Akeley's journeys, Ben Burbridge went looking for Mountain gorillas on two occasions. An American adventurer who had been on previous hunting safaris in East Africa, he came this time not to kill but instead to film and study. In the process, he hoped to clear up some misunderstandings regarding the gorilla: "to learn if he dances as does the chimpanzee; did he talk as certain authorities assert of monkeys. In a word, whether in him brutish or human characteristics predominate."[39] Burbridge himself thought them relatively harmless and wanted to capture some alive, an act that his porters considered crazy. After all, they knew stories of "women being stolen from their villages and kept captive in the forest by gorillas; how a warrior was killed while hunting, his heart torn out and carried away by a gorilla to make medicine; of a gorilla chief who was half man and lived in a distant jungle, to whom all the other gorillas paid tribute."[40]

After trekking across Tanganyika, Burbridge's party spent several weeks during the spring of 1923 searching for gorillas in the forest surrounding mounts Mikeno and Karisimbi. They saw plenty of spoor and nests and occasionally managed to get fleeting glimpses of the apes, long enough perhaps to make a kill but not to film. Burbridge then devised an array of strategies, such as imitating leopards and the distress cries of young gorillas to see if adults would respond. The favored one, he claimed, resulted from poor mathematical skills: "I discovered that if eight men went into a thicket and six of them came out and disappeared on the back trail, the gorillas would assume that all had left and would come to investigate. Then the camera would grind out its allotted plot."[41]

Encounters multiplied. One between Mikeno and Karisimbi illustrates Burbridge's flare for the dramatic:

> I could hear the light feet of other gorillas on the leaves. They were coming toward us stealthily. Shadows moved here and there; they seemed to multiply, merge, and separate. At first I thought it was my imagination, but they were really there, scores of them, standing silently where the shadows of overhanging branches hid them from distinct view.
>
> The lower bamboo swayed and bent as a young gorilla, perhaps eight years old, climbed like a Japanese acrobat. The stalk threatened to break under his hundred pounds, but he steadied himself upon another. At twenty-five feet he paused, and like a small boy peering over the fence at a ball game watched me turn the crank of my camera. His curiosity satisfied, he descended, biting at the bamboo branches in outraged fury and howling his disapproval at every step.
>
> His disappearance was the signal for an outburst of thunderous roars, followed by deep silence. I knew what was happening. The gorillas had paused to see what effect the din would have on us. It was the usual prologue to most of these weird performances.
>
> The squeal of a baby gorilla and the sudden crashing of the underbrush preceded a pandemonium of roars and chest beats as a gorilla weighing several hundred pounds, partially hidden by the leaf screen thirty feet distant, raged to and fro, smashing and tearing down everything in his path. Other gorillas concealed in the bamboo, their voices adding to the deafening tumult, grasped the stalks in each hand and shook them with such violence that their tops swayed and trembled like pipestems.
>
> I summoned every ounce of equanimity I possessed and stood peering through the latticed foliage, listening to the furious din of rattling teeth, roars, cheek and chest beats as the creatures menaced us, for all the world like some wild tribe preparing for battle. The outburst rose to a hellish zenith, and, as the weird echoes rang through the hills, flocks of parrots flew screaming across the sky, and from the valley below us came the distant trumpeting of the elephant herds.
>
> The demonstration having failed to rid them of the intruder, a sudden stillness fell upon the forest, during which I could see only the moving foliage and hear the click-click-click of my camera. Curiosity was fermenting beneath overhanging boughs; from the thicket came vague mutterings and chucklings and the shuffle of stealthy feet; then the forest sprang into life. Half a dozen gorillas showed themselves at once; a mother with her new-born infant upon her back; an old man framed by inquisitive wives, one leaning on the shoulder of another; but they could not be photographed in the sombre gloom of the forest. I called out, imitating the plaintive cry of a young gorilla caught by a leopard. This was our trump card to lure some of the females into the open to a supposed rescue.
>
> A screaming mob of gorillas formed in front of my camera—hesitated; then charging out of the forest, they paused terror-stricken before the one-eyed monster turned upon them, then broke and ran for cover.

Burbridge kept his Akeley camera rolling as a large male appeared in view:

> Again and again he roared his jungle war-cry while he beat his chest, each blow potent with crushing power sufficient to break a man's neck. . . . We met his charge with rapid fire, directed above his head. He paused in the midst of it, amazed at the thunderous reception, and, though unwounded, turned and fled, followed by the rest of his band.

Burbridge was now happy: "The gorilla would be taken back upon my film to the cities' millions. The wild man of the forest had been ensnared at last!"[42]

Next, they sought to capture a young gorilla. Burbridge had no intention of killing a family in order to get one. Instead, an attempt would be made to isolate a small juvenile and pin it to the ground manually, after which sacks would be used to subdue the captive. The Batwa wanted no part of any such thing, so the Bahutu porters would have to lend a hand. After a couple of missed opportunities, chance led to finding a very young male all alone. He proved to be an easy catch, and they called him Mikeno. The next one, a slightly larger young male, put up a valiant fight and bit one man badly before being subdued. He became known as Kivu. The final captures consisted of two small females, Lulanga and Quahalie, the latter being a misspelling of *kwa heri*, Kiswahili for "good-bye."

Before setting off for Dar es Salaam, the party spent time at the White Fathers mission station. While recuperating there, Burbridge described the gorillas' affection for one another as "touching," as they embraced "like little children who had not seen one another for a long time."[43] In November 1923, the party set off with the four gorillas in tow. Lulanga died of influenza on the way, followed by Kivu and Mikeno, the latter of which Burbridge said "held up his hands to me in a last farewell, on the shores of the Indian Ocean."[44] As per agreement, he personally delivered Quahalie to the Antwerp Zoological Gardens in December.

On his second trip during June and July 1925, Burbridge also had permission from Belgian authorities to capture four gorillas, again with the stipulation that one go to Antwerp. This time, he started from Leopoldville and followed a circuitous route to reach the country around Mount Mikeno. His first gorilla encounter came when a female and two young trailed behind the troop's silverback. While slowly tracking them, one of the young, a female, came close enough for Burbridge to pounce on and easily secure, as she weighed only twenty-two pounds. His next encounter proved to be far more difficult:

> I had grabbed a handful. Whether by accident or the instinct of gorilla-capturing, I clutched his throat and hung on with desperation born of the knowledge that

> I was battling for my life. The din of snarls and the thrashing of underbrush as we rolled over and over aroused my men to rush to my assistance. Twice I tore out of the gorilla's teeth and left a part of my clothing as a peace offering. Again and again I broke clutches that dragged my head and throat downward toward his open jaws. My gun-boy, racing through the jungle ahead of them all, flung himself into the fray. One after another piled on top of the young gorilla, who fought with the fury of a madman as he heaved and bucked under the weight of his enemies, refusing to accept defeat until spread-eagled and his hands and feet tied. While I lay gasping for breath, my men finally got him in a sack, and as this ripped from his attempts to escape, two more were slipped over him and tied.[45]

Burbridge, lucky for him, came away with only lacerations on his hands and a broken thumb. Some of his men also suffered minor injuries in the struggle and caging of their captive. He weighed 126 pounds, and they called him Bula Matadi, after the name the Congolese had given to Henry Morton Stanley in the 1880s. Its correct translation is "breaker of rocks," of which Burbridge was unaware.

Back at the camp, Bula Matadi raged at his captors, and for the first three days he refused food and water. Frustrated, Burbridge resorted to beating him with a padded club for acts of disobedience. Sometimes he used a firebrand when the club failed to do its job. The punishments eventually caused Bula Matadi to calm down, and a month after his capture the slow trek to the coast began. Late one night, driver ants invaded Bula Matadi's cage. He must have cried out, but no one claimed to have heard anything of the sort. In the morning, they found him dead with an uncounted number of bites on his body and "teeth fast locked."[46]

Despite not wanting to, Burbridge did shoot and kill one gorilla. While en route to Dar es Salaam, the party came across a group of females and young. Their cries alerted a silverback, and it charged. A single shot from Burbridge brought him down. Almost six feet in height, Burbridge claimed, "I was sorry, very sorry, for I did not want to destroy this magnificent creature who had fought for his young."[47] It was as though he had killed a man.

Burbridge did not describe the three other captures he made. Two of them survived. Marzo went to Antwerp, with Miss Congo winding up at the Jacksonville, Florida, home of Ben's brother Jim and his wife Juanita in October 1925. We will meet Miss Congo again in chapter 5.

Burbridge's 1926 movie *Gorilla Hunt* depicts the second journey and purports to be about a "huge brute" inhabiting the "jungles of Africa," one which "crushes skulls" and is the "King of the Beasts." Actually, most of the movie consists of the roundabout journey from Leopoldville to the country surrounding Mount Mikeno. Along the way, light is made of such things as Congolese dress, hairstyles, and dances in an effort to make the people look

Plate 4.2. Ben Burbridge and Miss Congo in Jacksonville, Florida. *Source:* Yerkes, "The Mind of a Gorilla."

childlike. Also true to Africa adventure films, lions and a huge tusked elephant are killed, and there's a meeting with supposed cannibals. Still, some interesting ethnographic images of Bambuti pygmies of the Ituri Forest and a Batutsi dance are on view. The gorilla hunt itself appears at the end of the movie and includes tracking around Mount Mikeno, the capture of the young female, and the killing of the large silverback. There's also a segment of what appears to be Miss Congo playfully unwinding film set out to dry. A restored version is available at the Museum of Modern Art in New York City.

Yale University, in conjunction with the Carnegie Institute, mounted an expedition in which no animals of any sort were to be hunted or captured. Instead, under the leadership of Professor Harold C. Bingham from Yale University's Department of Comparative Psychobiology, it would be an exploratory study of the habitats and living arrangements of gorillas within the Parc National Albert. During September and October 1929, they tracked troops in both the colder, higher zone of the volcanoes and their warmer lower slopes. Most often, this meant being content with following spoors and nests, with auditory contacts occurring on occasion. On several occasions, though, Bingham and his wife were able to study troops for short periods of time. As a result, only impressions and tentative conclusions could be offered, some of which proved to be accurate. For example, when examining several nests, they saw evidence of "listless departure in the morning," which seems to be the norm unless something startles a troop. They also noted that gorillas do not reoccupy nests, that sites are "secondary to other drives, for they sleep where night overtook them." Regarding diet, they found no evidence that gorillas were carnivorous or that they climbed trees for food.[48] They will, of course, do the latter.

A tragedy near the end did occur to mar what otherwise had been a reasonably successful outing for the expedition, given the circumstances. While following a troop, a silverback secreted himself under dense cover while the others moved on. Without warning, he charged, only to wheel about and go after a fleeing porter. Bingham was in a position to get a good view of the silverback and quickly got off a shot that brought him down but not before he stumbled on for forty paces.[49] It wasn't a pleasant moment for Bingham and, obviously, for the silverback.

MISTAKEN IDENTITIES

Other expeditions reported encountering Mountain gorillas when, in fact, they came across Eastern versions. This happened to the combined American Museum of Natural History and Columbia University one that left New York City on May 29, 1929. Under the leadership of William K. Gregory and Henry C. Raven, who were accompanied by J. H. McGregor and Earle T. Engle, it had the goal of collecting specimens for anatomical study. Plans called for an east-to-west crossing of Africa to secure an adult male and female each of Mountain and Western gorillas. The expedition reached Lake Kivu on August 4 and set up camp at Tschibinda in the mountains outside the Parc National near Bukavu. They first hired Bahutu guides, but, finding them ill suited to the task, Gregory turned to Batwa for help. As an experienced Africa hand, Raven would do all the shooting, using a 30-30-caliber Savage rifle and a 22-caliber rifle loaded with bullets containing small amounts of potassium cyanide. As he explained the hunt,

> Our two gorillas must be large adults and they must be shot only in the head. If shot through the body, many blood-vessels would be cut and it would be difficult to preserve the large body evenly; the preservative fluid injected would reach only certain regions; other regions would not be reached by it and would "go bad" perhaps on the ship going home.[50]

Only safe shots could thus be taken, which meant avoiding attempts in difficult terrain.

The first days brought signs of gorillas, but they didn't seem numerous on the ground. Raven's first attempt at a kill with the 22 failed. He did hit his subject, and when they saw him briefly the next day, he appeared none the worse for wear.[51] Soon thereafter came the first kill, a 460-pound male called the "Giant of Tschibinda" that required two direct hits to bring down. Then, after passing on what he thought was a male, Raven shot and killed a presumed female, only to discover it to be another male. Given the quota agreement, they wouldn't be getting a female gorilla here. As for the second kill, Gregory remarked,

> How clean and beautiful to us was the velvety black skin of his great face and how grave and calm the expression of his deep-set, brown eyes! Although it was indeed a pity to kill so noble a living monument of past ages, we had not murdered him wantonly and for sport. This gorilla was destined to be, though unconscious of it, a missionary of science.[52]

Both specimens were sent to Dar es Salaam for shipment to New York City, but the second arrived in poor condition due to an unclean head shot. It thus failed to serve as a "missionary of science." Although the study of gorillas wasn't on the expedition's agenda, Gregory did note,

> We had found these gorillas to be extremely shy, retiring animals, wanting nothing so much as to be let alone and repelling the advances of white busybodies. We had found them going in small roving bands of variable number, making their beds for the night under and in trees with spreading branches, and consuming a prodigious quantity of succulent vegetation.[53]

Given the location, the expedition had ventured into Eastern, not Mountain, gorilla territory.

On September 19, they left Tschibinda and, via a combination of boat, train, automobile, and marching, reached Yaoundé, Cameroon, on November 11. Finding gorillas in the dense surrounding forest proved harder than expected, and day after day went by without an opportunity to make a kill. The only gorillas seen were three young captives—one kept by a Greek merchant and two by a missionary. As the New Year approached, duties required both

Gregory and McGregor to leave for home. Engle had departed earlier, so it was left to Raven to continue the hunt. His first attempt to bring down a specimen yielded nothing but a trail of blood, and shortly thereafter sleeping sickness nearly killed him. The medical exam also turned up signs of malaria, hookworm, and roundworm infections.

After recovering sufficiently, Raven went out again, and by the time he left on January 5, 1931, he had managed to kill three adult males and collect several skeletons, ones likely the result of selling gorilla hands and skulls to wealthy customers abroad and the remains of a growing demand for bush meat, which brought good prices in towns and cities. According to Raven, events had moved in an unfavorable direction for Western gorillas in Cameroon:

> For centuries past gorillas and natives have been competitors. As the native population increased, new villages would be formed and more clearings made. Then epidemics would occur, killing off great numbers of natives, and their gardens would be neglected to run into second growth. The gorillas, with a constitution so nearly like that of man that they can find more food in human plantations than in the virgin forest, would move into these deserted clearings. There with an abundance of food they throve and congregated, to such an extent that eventually that if only a few natives remained they were actually driven out because of their inability to protect their crops against the gorillas. But with the advent of the white men's government, with the distribution of firearms, preventive medicine and the treatment for epidemic and infective diseases, man has the upper hand at present in this age-long struggle.[54]

In 1930, the American adventure photographers Martin and Osa Johnson appeared on the scene.[55] They had become international celebrities, hobnobbing with the likes of Jack London, Charlie Chaplin, and the Duke and Duchess of York, later king and queen. They had also become good friends with Carl Akeley, who convinced them to do a film on Africa. It didn't materialize until after his death, with the Bambuti pygmies of the Ituri Forest and Mountain gorillas the prime subjects of interest. Their only shots would be taken with cameras, for, as Martin noted, "There is nothing more disgusting . . . than the slaughter of animals for the sake of sport."[56]

The Johnsons arrived in the mountains in late 1930 at the head of a huge expedition comprised of 150 porters and two soundmen, intent on making the first talking film to come out of Africa. The stories told to them repeated some familiar refrains and added others. As Martin related in his book *Congorilla*,

> Gorilla guides themselves told me that when one of these animals attacks a man it tears off his arms and legs and tosses them away. One old timer related to me a vivid report of a gorilla attacking a native with a large club and beating him to

> death. Another story current in native circles is to the effect that when the leader of a gorilla pack becomes old, others in the group beat him to death. It is told also that when a leader gets too old to hold his grip, he goes away from his band and commits suicide. One native informed me that an old leader driven from the pack returns at night, kills each gorilla, one by one, and then commits suicide. . . . Other stories, varied and colorful, were in circulation, the most intriguing and exciting of which seemed to be the kidnapping of and cohabitation with human women by these hairy apes.[57]

He went on to say that all this must be taken as "tongue in cheek."

Their own accounts contained a mixture of both old images and new, with the first encounter on Mount Mikeno reading much like Du Chaillu:

> Suddenly, a huge black face peered through the bamboo. It was then that I found the source for the fabulous tales about this fearsome looking beast. The face was black, like oiled and polished leather; black as anything you will ever see. Framing it was black, close cropped hair through which round, small ears were peeping. Two eyes stared solemnly and directly at me. There was something about those eyes that suggested an evil spirit. They seemed to be glaring right through me as though some Satanic judge of the nether world were considering the penalty for one who dared invade his hidden precincts. No wonder people believed these hairy creatures to be half man, half demon. That face, with its curling, sneering lips, looked cold, cruel and murderous.
>
> Only a few seconds I looked into those stolid, musing eyes, and in the twinkling of an eyelash the head vanished. Then, suddenly, the stillness of the mountain side was broken with an ear splitting scream that ripped through the jungle. It was followed by another and still another in a nerve-racking blood chilling chorus. These shrieking cries went through, around and over us, only to hit a mountain side that hurled them back with painful echoes caused by the roaring assault. It seemed that the demons of hell had broken through the earth's crust, ready to tear it apart and hurl the pieces into the universe.
>
> The surprise of this horrible medley left me rigid and it was several minutes before I regained my poise. Then, recovering my eagerness to picture these apes, I ran to another screaming bush. I did not have time to set up the camera, but I was quick enough to see another black shadow fading into dense bamboo. Believing there was no chance for a picture, I set the camera down and stepped closer to the clump. Then, not more than fifteen feet away, an enormous gorilla slowly rose on his legs, grasping vines with two black hands. He opened his enormous mouth and aimed the most blood curdling yell ever heard, directly at me. I could see his red tongue and blood-red gums. Sword-like fangs were bared by the snarling lips; flanking them were teeth, huge and sharp. Had I not known better, I would have sworn that this ape was ten feet tall and weighed a thousand pounds, so vivid was the impression. . . .
>
> As I stared at that horrid face my legs were locked with bonds of terror. I had no gun or other weapon for protection and into my mind leaped all of the fearful

> gorilla stories I had ever heard. I expected at any minute to be torn limb from limb. When the ape whirled around, dropped on all fours and ran in the opposite direction, the spell was broken and the warm blood coursed through my veins.[58]

For the most part, though, the Johnsons played hide-and-seek with gorilla families, as poor lighting and dense vegetation made prolonged contacts difficult.

From Virunga country, the party moved into the mountains south of the Parc, where they discovered gorillas in abundance. Martin thought the area might hold as many as 20,000 individuals. Although the people here displayed a "healthy fear" of the apes, no one knew anything about attacks on humans having taken place. They also thought the idea of gorillas capturing women "absurd." By this time, the couple had concluded that gorillas were not aggressive unless provoked by humans.

In their analysis of the available evidence, primate specialists Robert M. and Ada W. Yerkes came to the same conclusion, noting that in general gorillas were "more often shy and retiring than annoyingly aggressive or inquisitive" and that "no adequate evidence" existed of a "natural hostility toward man."[59]

Plate 4.3. Ingagi and Congo after being captured by Martin and Osa Johnson. *Source: Congorilla.*

Belgian authorities had given the Johnsons permission to capture a young gorilla, and one day they came across a family. Two young ones scurried up a tree, and after driving off the silverback trying to protect them, the tree was cut down, making the capture of the dazed infants easy. Both turned out to be males about five years old, one receiving the name Ingagi, the other Congo. Just as the Johnsons were preparing to leave for Nairobi, a group of local men came by carrying a very sick young gorilla. Although not optimistic about its survival, the Johnsons bought him anyway. Okero, or Snowball (he wasn't possessed of white hair) to some, did survive, and the three young gorillas were used to simulate action before the cameras that the crew hadn't been able to get in the wild. Okero wound up at the National Zoological Park in Washington, D.C., where he survived for just over a year. As we shall soon see, Congo and Ingagi fared much better at the San Diego Zoo. Still, Martin lamented what he did and vowed never again to "send another into captivity."[60]

The movie version of *Congorilla* debuted in 1932. The segment on gorillas comes at the end, with the capture of Ingagi and Congo and the rescue of Okero featured. The narration stresses the dangers involved in tracking gorillas and in a departure from the book refers to them as "brutes" and "beasts" in an obvious attempt to play on what most audiences would think gorillas to be.

Despite being labeled Mountain gorillas, the three captives were taken from the same general area as those of the McGregor and Raven expedition and, consequently, had to be Eastern. Indeed, this seems to have been the case for most of the gorillas the Johnsons encountered. A revealing bit of evidence is provided by the great number of them seen during the party's earlier trek through country west of Mount Mikeno, numbers so large that Martin felt they faced no danger whatsoever of extinction. And when the party reached true Mountain gorilla country in order to visit Carl Akeley's grave, nary a one was seen.

The Mountain gorilla also attracted the attention of the Italian adventurer Attilo Gatti, who claimed he wanted to become "a famous explorer and a gorilla hunter extraordinary" after having first heard of them at age seven.[61] The need to recover from an illness suffered during World War I put a hold on this ambition, but eventually Gatti did begin a hunting and collecting career in Africa. During one such outing, he received official permission to enter the Tschibinda forest south of Lake Kivu to obtain a gorilla specimen for the Royal Natural Museum of Florence. While on the hunt, he told of spying a family nearby. A Batwa guide, whom he mistakenly called a Mbuti, gave a leopard call to spook them, which brought a large male forward called Moami Ngagi, or King Gorilla, which, according to local accounts, had crushed two men to death a few years earlier and then went after several others in the group.[62] Speared, the male ran off, only to reappear several months later fully recovered. Now, as Gatti put it,

> Wild with hate Moami closed his terrible fingers around a tree which ten robust men could scarcely have broken, and snapped the trunk in two with a crackling sound like a cannon shot. Brandishing this formidable weapon he advanced a few steps, emitting a howl of a very different character. It must have been an order, for the females and the young ones promptly dropped down from the trees and trotted away toward the bushes. As soon as they had disappeared, Moami in a last burst of rage seized the trunk of the tree with his fangs, tore it to pieces as though it were a stalk of soft sugar cane, and threw it far away.[63]

Moami then left to rejoin his group, and, in pursuit, Gatti said,

> Instantly three horrible howls sounded in my ears and three enormous bodies bounced out of the thicket, Moami Ngagi in the lead, waving a huge arm and beating his breast in a rage. His almost human face was disfigured by a cruel grimace, the ferocious eyes shining like two lumps of coal, the mouth stretched open in a snarl displaying long, yellow fangs.[64]

Gatti fired but to no effect as Moami continued to come forward. A second shot did the job, with the animal reportedly brushing the hunter's knees as it fell. Then two females came after him. The gun jammed, and after getting it unjammed, Gatti claimed to have remembered the threat of a 20,000-franc fine and possibly a jail sentence if he killed more than one gorilla. So he shot one in the arm, followed by two more shots in the air, sending both of them on their ways. Obviously, Gatti either fabricated or embellished the story for book sales—he was a prolific author of popular adventure books. There is a photo, though, in his *Tom-Toms in the Night* showing Gatti posed next to a male said to have weighed 482 pounds and purportedly Moami, although that's not specified. The caption notes it having been sent to Witwatersrand University in South Africa, where renowned paleoanthropologist Raymond Dart had become interested in gorillas.[65]

In order to reach a larger audience, Gatti retold the harrowing Du Chaillu–like story of Moami Ngagi for the trade magazines *Popular Mechanics* and *Field & Stream*.[66] In both, he sought to emphasize the beast image of gorillas, saying that, from the deep bush,

> almost instantly there followed an indescribable, nerve-racking howl that sounded like a canon shot. It seemed to be a combination of the roar of a lion, the pitiful yowl of a dog in agony and the cry of a mortally wounded man. Never, in all Africa had I ever heard such an awesome and startling cry. It bespoke rage, fury, power and danger.

Plate 4.4. The realization of Carl Akeley's Gorilla Diorama at the American Museum of Natural History. *Source:* Gorilla Group in Africa Hall, American Museum of Natural History, image 315491.

This time, purportedly two shots to the heart brought down what Gatti called a "very dangerous beast, probably the most dangerous of all African animals," one that he now said measured six feet eight inches in height and weighed 530 pounds. In the longer *Field & Stream* piece, he noted that after killing Moami Ngagi, before him stood a half-grown male and two females "waist-deep in the tangle of brush, they were glowering at me with a maddened aspect on their grotesque faces, and bellowing unholy sounds in a manner that assured me they need but the slightest encouragement to continue the charge." Gatti moved slightly, which brought the largest female menacingly at him. Managing to expel the jammed cartridge, he quickly reloaded and fired. A howl came from her, and three gorillas retreated, the hunt coming to an end.

Attilo Gatti represents the last of a dying breed, the Great White Hunter hell-bent on shooting Mountain gorillas. Symbolic of the changing times was the opening of the Akeley African Hall exhibit at the American Museum of Natural History on May 19, 1936. James L. Clark, a former student and colleague of Akeley, had augmented the collection and oversaw completion of the hall in his position as director of arts, preparation, and installation.[67] One of the dioramas featured a Mountain gorilla family, with the Karisimbi silverback standing majestically against a volcanic background. It is still there today, showing a scene of natural beauty and serenity to all who have and will pass by for a view and a reminder that the gorillas paid with their lives to be there on exhibit.

By the 1930s, a number of people, through books, films, and exhibits, had reported on the existence of Mountain gorillas, and using this information, along with his own observations, Uganda game warden Captain Charles R. S. Pitman attempted to summarize what was known about them at the time.[68] While the results indicate that much had been learned to challenge prevailing myths and suppositions, especially about the dangers to humans posed by gorillas, more would be needed if they were to be protected and their habitats preserved. Delia Akeley actually hit the nail on the head by remarking in 1930, "The real truth concerning the habits and characteristics of apes and monkeys can be learned only through exhaustive study. Years must be spent by the student in the lonely forests where the animals live."[69] But such field studies would take a while to materialize, as World War II loomed on the horizon.

NOTES

1. John Hanning Speke, *Journal of the Discovery of the Nile* (New York: Harper & Brothers Publishers, 1864), 226.
2. Verney Lovett Cameron, *Across Africa* (London: George Philip & Co., 1885), 221.
3. George B. Schaller, *The Year of the Gorilla* (Chicago: University of Chicago Press, 1864), 53.
4. Pascal Sicotte and Prosper Uwengeli, "Reflections on the Concept of Nature and Gorillas in Rwanda: Implications for Conservation," in *Primates Face to Face: The Conservation Implications of Human-Nonhuman Interconnections*, ed. Augustin Fuentes and Linda D. Wolfe (Cambridge: Cambridge University Press, 2002), 163–81.
5. Ewart S. Grogan and Arthur H. Sharp, *From the Cape to Cairo*, 2nd ed. (London: Thomas Nelson & Sons, 1920), 195.
6. George B. Schaller, *The Mountain Gorilla: Ecology and Behavior* (Chicago: University of Chicago Press, 1963), 390, translated from the German by Dennis McCort.
7. Paul Matschie, "Uber einen Gorilla aus Deutsch-Ostafrika," *Sitzungsberichte der Gesellschaft Naturforschender Freunde* 1903: 253–59.
8. P. J. Pretorius, *Jungle Man, the Autobiography of Major P. J. Pretorius* (New York: E. P. Dutton & Company, 1948), 119–20.
9. Ibid., 121.
10. Thomas Alexander Barns, *Across the Great Craterland to the Congo* (New York: Alfred A. Knopf, 1924), 145.
11. Thomas Alexander Barns, *The Wonderland of the Eastern Congo* (London: G. P. Putnam's Sons, 1922), 87.

12. Barns, *Across the Great Craterland to the Congo*, 135.

13. Ibid., 151.

14. Prince William of Sweden, *Among Pygmies and Gorillas* (New York: E. P. Dutton and Company, n.d.), 8–9.

15. Ibid., 195.

16. For an overview of Ward's work, see Kristy Breedon, "Herbert Ward: Sculpture in the Circum-Atlantic World," *Visual Culture in Britain*, Special Issue, "British Culture c. 1757–1947: Global Context" 11 (2010): 265–83.

17. For the details of Carl Akeley's life, see Mary L. Jobe Akeley, *The Wilderness Lives Again: Carl Akeley and the Great Adventure* (New York: Dodd, Mead & Company, 1940); Penelope Bodry-Sanders, *Carl Akeley: Africa's Collector, Africa's Savior* (St. Paul, MN: Paragon House Publisher, 1991); and Jay Kirk, *Kingdom under Glass: A Tale of Obsession, Adventure, and One Man's Quest to Preserve the World's Great Animals* (New York: Henry Holt and Company, 2010). An interesting although at times dense social theory look at Akeley can be found in Jeanette Eileen Jones, "'Gorilla Trails in Paradise': Carl Akeley, Mary Bradley, and the American Search for the Missing Link," *Journal of American Culture* 29 (2006): 321–36.

18. Carl Akeley, *In Brightest Africa* (Garden City, NY: Doubleday, Page & Company, 1923), 8.

19. Jumbo had become a favorite of the British public, and his sale by the London Zoological Society to an American circus provoked widespread outrage.

20. Akeley, *In Brightest Africa*, 114.

21. Delia J. Akeley, *"J. T. Jr.": The Biography of an African Monkey* (New York: The Macmillan Company, 1928).

22. Akeley, *In Brightest Africa*, 196.

23. Mary Hastings Bradley, *On the Gorilla Trail* (New York: D. Appleton and Company, 1922), 3.

24. Ibid., 75.

25. Akeley, *In Brightest Africa*, 215.

26. Carl Akeley, "Gorillas—Real and Mythical," *Natural History* 23 (1923): 430.

27. Akeley, *In Brightest Africa*, 217.

28. Ibid., 223.

29. Bradley, *On the Gorilla Trail*, 116.

30. Ibid., 121.

31. Akeley, *In Brightest Africa*, 235.

32. Bradley, *On the Gorilla Trail*, 123.

33. Akeley, "Gorillas—Real and Mythical," 447.

34. Ibid., 430.

35. Anonymous, *Traveller's Guide to the Belgian Congo and the Ruanda-Urundi*, 2nd ed. (Brussels: Tourist Bureau for the Belgian Congo and Ruanda-Urundi, 1956), 151.

36. For a brief biography of Delia, see Elizabeth Fagg Olds, *Women of the Four Winds* (Boston: Houghton Mifflin Company, 1985), 71–153.

37. Mary L. Jobe Akeley, *Carl Akeley's Africa* (New York: Dodd, Mead & Company, 1929), 146, emphasis in the original.

38. J. M. Derscheid, "Notes sur les Gorillas des Volcans du Kivu (Parc National Albert)," *Extrait des Annales de la Societé Royale de Belgique* 58 (1927): 149–59.

39. Ben Burbridge, *Gorilla: Tracking and Capturing the Ape-Man of Africa* (New York: The Century Company, 1928), 210.

40. Ibid., 227.

41. Ibid., 230.

42. Ibid., 235–39.

43. Ibid., 271.

44. Ibid., 277.

45. Ibid., 256–57.

46. Ibid., 265.

47. Ibid., 267.

48. Harold C. Bingham, *Gorillas in a Native Habitat* (Washington, DC: Carnegie Institute of Washington, 1932), 29–35.

49. Ibid., 52–57.

50. W. K. Gregory and H. C. Raven, *In Quest of Gorillas* (New Bedford, MA: Darwin Press, 1937), 78.

51. H. C. Raven, "Gorilla: The Greatest of All Apes," *Natural History* 31 (1931): 231–42.

52. Gregory and Raven, *In Quest of Gorillas*, 91.

53. Ibid., 109.

54. Ibid., 229.

55. Pascal J. Imperato and Elanore M. Imperato, *They Married Adventure: The Wandering Lives of Marin and Osa Johnson* (New Brunswick, NJ: Rutgers University Press, 1992), is a well-researched and readable account of their adventures together.

56. Martin E. Johnson, *Camera Trails in Africa* (New York: Grosset & Dunlap, 1924), 8.

57. Martin E. Johnson, *Congorilla* (New York: Brewer, Warren & Putnam, 1931), 112–13.

58. Ibid., 132–34.

59. Robert M. Yerkes and Ada W. Yerkes, *The Great Apes: A Study of Anthropoid Life* (New Haven, CT: Yale University Press, 1929), 455.

60. Johnson, *Congorilla*, 281.

61. Attilo Gatti, *Tom-Toms in the Night* (London: Hutchinson & Co., 1932), 101.

62. The Batwa are renowned hunters and trackers. They, along with the Bahutu and Batutsi, are encompassed within a larger Banyarwanda polity, the origins of which date back to the fourteenth century. All speak Kirundi. While the status of the Batwa is largely unchanging, there's considerable movement between Bahutu and Batutsi, depending mostly on marriage and economic circumstances, with royal patronage and large herds of cattle the traditional signifiers of Batutsi status.

63. Gatti, *Tom-Toms in the Night*, 132.

64. Ibid., 140.

65. Ibid., opposite 108.

66. Attilo Gatti, "Among the Pygmies and Gorillas," *Popular Mechanics*, September 1932, 418–21, 118A, and "Gorilla," *Field & Stream*, October 1932, 18–20, 66–67, 73.

67. James L. Clark, *Good Hunting, Fifty Years of Collecting and Preparing Habitat Groups for the American Museum* (Norman: University of Oklahoma Press, 1966).

68. Charles R. S. Pitman, *A Game Warden among His Charges* (London: Nisbet & Co., 1931), chaps. 10 and 11.

69. Delia Akeley, *Jungle Portraits* (New York: The Macmillan Company, 1930), 28.

Chapter Five

Knowledge Comes to the Rescue

FALSE STARTS

Besides the expedition headed by Gregory and Raven, only one other with either scientific or collecting objectives targeted Western gorillas in the 1930s. Headed by multimillionaire George Vanderbilt, it marched westward from Mombasa in 1934 destined for Cameroon, intent on collecting specimens for the Africa exhibit at the Academy of Natural Sciences of Philadelphia. At the completion of the journey, 400 mammals, 1,294 birds, 400 reptiles and amphibians, 3,000 fish, several hundred mollusks, and 16,670 insects were secured.[1] For unspecified reasons, the expedition bypassed the Virungas and adjacent highlands and instead ventured through the Ituri Forest, encountering no signs of gorilla presence anywhere along the route. Eventually, four Western males killed by locals were acquired from the area surrounding the Sanaga River in French Equatorial Africa.[2] At no time did members of the expedition make an attempt to study gorillas in the field, and thus added little in the way of knowledge about them.

The Belgian filmmaker and adventurer Armand Denis appears to have mounted the only serious effort by a non-African to locate gorillas during World War II. He had established a breeding program for chimpanzees in Florida and wanted to do the same for gorillas. Consequently, when word reached him about a large number of them living in French Equatorial Africa, a claim discounted by most authorities at the time, Denis decided to risk the dangerous Atlantic crossing aboard a Norwegian freighter in February 1944. It got him safely to Monrovia, Liberia, and from there he ventured overland, mostly by hitchhiking, to Brazzaville.[3] Heading north from the city through largely uncharted territory, Denis reached a place where people knew about gorillas because they hunted them for food. The first live one presented to him turned out to be badly diseased, so he had it mercifully put down. Next came a large male killed by villagers that provided Denis with an opportunity to do some close-up physical observations and take measurements. At the sight of the victim, he vowed to do something about the killing of gorillas.

Shortly thereafter, the villagers presented Denis with a young, live male. A week later, nine more youngsters appeared on his doorstep, including an infant from a hunt that resulted in seven of the family being killed. More hunts followed, and over the course of two months Denis acquired the thirty gorillas authorities in Brazzaville allocated to him for export. He could do nothing, of course, about the killings other than to console himself with the thought of having saved thirty gorillas from such a fate. Denis filmed the hunts and also took some footage of gorillas in the wild, seemingly a first for the Western version. In the process, he confirmed that gorillas used cupped hands, not fists, to beat their chests.

Miraculously, Denis and his party returned to Brazzaville with thirty live gorillas in tow. A good-sized male replaced a female who had died before the journey began. Suddenly, though, things started to go bad, and within a week sixteen of the gorillas had died from an unknown disease. The remaining fourteen did survive the trip to the port of Matadi, only to die there while waiting for the arrival of a ship headed to the United States. On top of this, Denis lost his film when the ship carrying it sank off Bermuda—so much time, money, and lives wasted for nothing.

With no new information of significance at its disposal, the world was still stuck with two competing narratives about gorillas: the "beastly" one based on Du Chaillu and what might be called that of the "gentle giant" one, derived from Carl Akeley and others who had encountered Mountain gorillas. In 1950, Lucien Blancou went to equatorial Africa hoping to separate fact from fiction and thus resolve the matter once and for all. He didn't, however, make much headway because of an inability to find groups to study in detail, even within protected areas. This led him to conclude that gorillas were not usually aggressive toward humans, preferring instead to stay away from them when at all possible. Blancou also reported on a tourist trade in skulls and the existence of a "war" between humans and gorillas then in progress because of increasing competition for land. While their numbers did seem to be declining, he thought that Western gorillas were not faced with extinction, at least not anytime soon.[4]

Also in 1950, Columbus, Ohio, native William Presley Said arrived in Gabon bent on capturing gorillas for profit. The University of Wisconsin–Madison had agreed to help cover his costs and pay $2,000 per animal for research purposes. Said's initial effort proved to be futile, and it cost him a severe attack of malaria. Once recovered, and undeterred, he set off again with a large party of Bacola hunters to find a suitable troop. This time, success came their way, and after the silverback and seven other adults had been speared to death, Said captured a young one with his bare hands. A trek with Mbeti hunters into the forest near Boma in the Belgian Congo followed. It yielded five more young gorillas once the adults and adolescents had been either killed or driven off. Two of them failed to survive, leaving Said with a total of four to take back to the United States. When he arrived in New York City, he found out that the University of Wisconsin had canceled its research plans due to a lack of funds. Said then offered to sell the four gorillas to the Columbus Zoo for $10,000.[5] It agreed to take two of them, which we'll meet in chapter 6. Another, called Bobo, wound up being purchased by Bill Lowman, a fisherman in Washington State. Bobo survived the long drive from Columbus and stayed with the Lowman family until 1953, when Seattle's Woodland Park Zoo took over his care. Bobo became a big hit with visitors until unexpectedly dying on February 22, 1968. He'll appear again in chapter 6.

The fourth went to Noell's Ark Gorilla Show, one of those traveling animal extravaganzas that crisscrossed the United States in the mid-twentieth century.[6] Despite the name, Noell's, which was in business from 1940 to 1971, featured chimpanzees, the most popular acts being boxing and wrestling matches with volunteers from the audience. To avoid serious accidents, the chimpanzees wore muzzles and gloves, while the humans, who never bested their simian foes, donned helmets. Gorillas didn't join the show until the 1950s. The first three that Noell's purchased or tried to purchase were sick and died almost immediately. The one from Said survived for a couple of years, delighting onlookers with various antics, although not of the wrestling and boxing kind, before a parasitic infection claimed his life.

Said went back to equatorial Africa to capture more young gorillas to send to zoos in Chicago, Cincinnati, Toronto, and Brazzaville. In the process, he gained the nickname "Gorilla Bill" and wound up being featured in a *Life* magazine story called "*Life* Goes on a Gorilla Hunt."

Henry Geddes next arrived on the scene. A kind of jack-of-all-trades, by this point in his life he had settled on a career in film and one day, without any prior knowledge of place or subject, decided to photograph Western gorillas. At Leopoldville in the Belgian Congo, Geddes teamed up with the American film director and double/stuntman Yakima Canutt, at the time on the lookout for scenes of gorillas to include in an upcoming film. During a stop in Brazzaville, eight more Europeans and a cadre of African workers joined up for a journey to find the best location in which to film. All of them agreed not to kill a gorilla unless absolutely necessary, something that made no sense to Africans met along the way, who repeated the same old story of gorillas carrying off women. People in this part of Africa also hunted them for food.

After many weeks on the road, Geddes settled on a site in Moyen Congo (today's Republic of the Congo) near the border with Gabon.

The plan called for encircling a family with a wire-mesh net and then driving its members into a clearing where a large metal cage awaited with pineapples and tender shoots inside in hopes that a gorilla would be tempted to enter and trigger the door to slam shut. With around 450 Mbeti men to serve as drivers and untold numbers of women and other followers, the expedition became by far the largest ever assembled to search for gorillas. After numerous failures, they did find a family to surround with the net. At first, none of its members cooperated by entering the clearing despite the bounty of food present and the use of an implement to simulate the sound of thunder, which gorillas aren't fond of hearing. Eventually, though, a young male did take the bait and walked into the cage, only to break out with relative ease. Determined to get a gorilla, Canutt, who had been a rodeo cowboy before entering the film industry, managed to lasso another young male by the arm, and several Mbeti men joined in to secure it with a rope net. Thus subdued, they had a healthy live specimen to send to the Brazzaville Zoo.

After ten days of humans treading back and forth, the vegetation surrounding the clearing had been matted down sufficiently for photos of the other gorillas to be taken. Prior to that, they got only a few shots of a silverback repeatedly charging the metal fence. The family then went on its way minus the one member but with none killed or physically harmed at the site.

Geddes wrote a book called simply *Gorilla*. It includes a number of photos, although only a few of them depicting gorillas were actually taken by the expedition. In fact, more photos of young, bare-breasted women wound up being put on display for readers to ogle. As far as the story goes, gorillas really didn't enter significantly until the very end, when the hunt began. And then it mostly revolved around the silverback's (here called a *grand garçon*) demeanor as he reacted to the drive and the fence. Instead, most of the pages were filled with the adventures and problems of the expedition. In truth, it's hard to tell how much of the narrative is fiction, embellishment, or fact. Nonetheless, the story is sympathetic in tone, emphasizing the gorillas' courage and human qualities as opposed to beastly ones.

No movie per se of the expedition ever appeared, although some of Yakima Canutt's shots showed up in the 1953 potboiler *Mogambo*, starring Clark Gable, Ava Gardner, and Grace Kelly. The geography is all mixed up as the plot alternates between widely separated settings in Uganda, Kenya, and French Equatorial Africa that are made to look close by one another. In addition, a large male is shown being killed during a charge, an event even more graphically depicted on the movie's advertising poster, where a huge, snarling beast with mighty arms raised is set to attack a fearless white hunter poised to bring him down. As noted, no killing ever happened during the Geddes expedition.

Plate 5.1. One of the posters advertising the movie *Mogambo*. *Source:* Photofest.

Mogambo, though, did cost a life, that of "Gorilla Bill," in April 1952. Hired as a consultant for the film, the truck he was driving mysteriously crashed on the way to pick up the cast.

Killings and captures had also begun taking place among populations of Eastern gorillas. A Belgian mine company worker reportedly killed nine of them, and in 1948 sixty adults were sacrificed in order to capture eleven young, of which only one survived. Charles Cordier, not the famous sculptor but rather a Swiss American animal collector, specialized in the capture of gorillas. In 1957, he secured a gorilla family for inclusion in the 1960 movie *Les seigneurs de la forêt* (an English version is titled *Masters of the Congo Jungle*, dramatically narrated by Orson Welles and William Warfield). Cordier also trapped an array of Eastern gorillas to ship to zoos, including those in Antwerp, London, the Bronx, and San Diego.[7] The method that Cordier developed consisted of driving his subject into netting, surrounding the net with poles to prevent escape and avoid injury, and then slowly releasing the poles in the direction of an awaiting cage. This allowed him to capture unharmed gorillas older than infants.[8]

TRAVELLERS REST

The turn to a more systematic approach to the study of gorillas began in the mid-1950s, with M. Walter Baumgartel being the unlikely initial source. German by birth, he had lived in South Africa for many years and flew for the country's air force as an aerial photographer during World War II. Desirous of returning to England, he sold his share of a publishing business in Johannesburg and on the way out visited East Africa, finishing up in Uganda for a trip down the Nile to Egypt. Reenergized about Africa, Baumgartel couldn't resist responding to a newspaper ad that read "Partner Wanted for Hotel Project in Western Uganda," and in March 1955 he arrived at an innlike establishment called Travellers Rest, just outside Kisoro, the northern entryway to the Virunga volcanoes. While marveling at the scenery, Baumgartel said that his spirits "took a nose dive" because the inn consisted of just the owner's private residence and a small group of "miserable huts."[9] Still, he forged ahead, thinking a more strategic place couldn't be found, as it stood at the crossroads leading to Rwanda and the Belgian Congo.

Dealing with his partner and the owners of the inn tried Baumgartel's patience and emptied his pocketbook, but before the year ended he found himself in sole possession of the ramshackle establishment, hoping that he could make it into a place to serve as a base both for studying gorillas and for high-roller tourists who wanted to see them. The existence of gorillas in the nearby mountains hadn't been fully confirmed until 1929. He soon realized that improving and running the property would consume most of his time, and thus Baumgartel contacted famed paleoanthropologist Louis Leakey about obtaining an assistant to begin the studies. A few weeks later, Leakey wrote saying that he knew just the person, Rosalie Osborn, his secretary and onetime lover, at the Coryndon Museum in Nairobi. She jumped at the opportunity and set off for Travellers Rest in October 1956 to begin research under Baumgartel's direction. As matters turned out, Osborn spent only forty-two days tracking the apes to little or no avail due to the difficulties of finding them in the dense vegetative cover. They tried putting out food for the gorillas, hoping that it would cause them to stay nearby. When this didn't work, Osborn decided to call it quits and go home to England, where she took a job at the British Museum of Natural History in London's Kensington district. Osborn did, however, agree to remain at Travellers Rest until a successor arrived. Her published report appeared six years later and included only superficial observations on gorilla foods, movements, locomotion, vocalizations, and nests, she herself commenting that the notes were "insufficient material from which to make generalizations and conclusions."[10] Osborn may, however, have been the first researcher to see an act of copulation by free-ranging gorillas, although at the time it appeared to her as just playful behavior. Shortly thereafter, Baumgartel and his ace tracker and master guide Rubin Rwanzagire filled the void by witnessing two acts.[11]

Osborn's successor, Jill Donisthorpe, had somewhat greater success. A young woman with a travel bug, one day while living in Nairobi she saw a newspaper ad placed by Baumgartel seeking a scientist to do "research on apes."[12] She had earned a BSc degree from Bristol University, and her interest in gorillas surfaced after meeting Raymond Dart, who along with Philip Tobias had established the Uganda Gorilla Research Unit at Witwatersrand University in South Africa. She thus decided to apply for the position and shortly afterward found herself at Travellers Rest to pick up from where Osborn had left off. Although also precluded from doing an in-depth study because of the difficult environmental circumstances and the wandering habits of gorillas, Donisthorpe did produce a number of observations during 122 days in the field between February and September 1957 for others to build upon. For example, she noted that the gorillas seemed to live mostly between 7,500 and 9,500 feet in elevation and, contrary to the prevailing view, frequently built nests in trees. She also recorded a variety of gorilla sounds and collected specimens of twenty-two plants eaten by them.[13] While Donisthorpe, too, published only one scientific report, she wrote two shorter pieces for general audiences and a rather charming although little-known book about her gorilla experiences called *Gorilla Mountain*.[14]

After obtaining a college degree, Rosalie Osborn returned to East Africa not to study gorillas but rather to teach children. Jill Donisthorpe went back many times during a long life of globetrotting, but the field experience caused her to lose interest in gorillas. As she noted, "I didn't like them, and they didn't like me."[15]

Other visitors to the area included Oliver Milton, a former game ranger in Tanganyika, and John Blower, then a game ranger in Uganda. They hoped to photograph Mountain gorillas since this had yet to happen here.[16] Luck, however, bypassed them, as their only sighting turned out to be a brief glimpse of a lone gorilla. What Milton and Blower did record was a forest boundary in rapid retreat due to the expansion of human settlements upslope into a setting known to locals as *Rwengaji*, or the "place of the gorillas." The whole area, they concluded, should be made a national park where no one would be allowed to enter without special permission and from which all cattle should be strictly forbidden. Otherwise, they felt that the remaining gorillas would either die off or flee into Rwanda and the Belgian Congo.

In 1958, the Japanese Monkey Centre, located just outside Nagoya, sent Kinji Imanishi and Junishiro Itani on an expedition to see about obtaining gorillas for a Monkey Zoo, designed to serve as a center for primate study and public education.[17] They made Travellers Rest their first stop before going on to survey the areas inhabited by Western gorillas. Because of less human interference and comparative ease of access, the Mountain gorilla seemed the best choice for a study, and the following year a research team from the Centre arrived. It tracked a number of groups, but the study of sociological behavior in the manner the Centre had done with Japanese monkeys proved to be impossible, again because of environmental difficulties. As a result, their report did not add much to what Bingham and Donisthorpe had found.[18] The Centre then decided to try the strategy of putting out foods to attract gorillas to specific sites where they could be readily observed. As with the earlier efforts by Baumgartel and Osborn, the apes paid the foods no mind. Given this and tracking difficulties, the Centre decided to end the study.

Without bothering to get permission, Baumgartel set about turning Travellers Rest into a business focused on gorilla tourism. When the Ugandan authorities got wind of this, they ordered him to cease and desist. He did as told, but his persistent efforts convinced the right people that money from tourists would help aid conservation efforts. In return, Baumgartel became an honorary game warden. And just as he predicted, tourists did start to come in increasing numbers but only under strict provisions. All of them had to get permission from the game warden in Entebbe, and Baumgartel usually allowed only two people at a time to go in search of gorillas.[19]

Not all visitors, however, achieved their heart's desires. Two who did were Ken Newman and his wife. Led by Rwanzagire, they watched an old male that Baumgartel had named Saza Chief eating contentedly in the company of two females and a young male. Upon finishing their meal, they slowly walked away, Saza, according to Ken, showing "how concerned he was for his family, constantly looking around to ascertain their whereabouts."[20]

Another visitor was Paul A. Zahl from the National Geographic Society. He, too, had success, and, although he knew that gorillas were not savage beasts, he made them seem so for readers of a magazine that specialized in stories laced with scenes and tales of danger in exotic settings. For instance, an encounter led him to remark, "With nostrils flared and lip curled back to expose fearsome white teeth, the creature stopped short some 20 feet away. Sunken eyes, the essence of savage malevolence, fastened on us." To make his point even more graphic, Zahl called the gorilla a "400 pound nightmare."[21]

For years, stories had circulated about leopards killing gorillas, a contention supported by Charles Pitman in his 1931 summary of knowledge about gorillas. Using hair seen in leopard droppings, he concluded that they "exact a heavy toll."[22] Still, no such incident had ever been officially recorded. This changed one day when Rwazangire saw a leopard snatch and kill a young male, and later he came upon the same leopard eating a young female. Afterward, he and Baumgartel discovered the body of a third gorilla seemingly killed in the same manner. Spoor also showed that the leopard tracked the rest of the family as it retreated to the Rwanda side of the Virungas.[23] The two men thus had to give up following the trail. It's now clear that killings by leopards are extremely rare and probably restricted to specific ones that have developed a taste for gorillas and are willing to take the risk to satisfy it.

Baumgartel sold Travellers Rest to a Swiss group that in turn sold it to an Indian entrepreneur who then had it confiscated by the Ugandan government when Indians were ordered to leave the country in the late 1960s. Further ownership changes followed, and in 1999 the facility underwent major renovations and now serves as the main stopping point for visitors to Mgahinga National Park, established in 1991.

YEAR OF THE GORILLA

The first truly in-depth study of gorillas began with George B. Schaller. Born in Germany in 1933, Schaller's family survived the horrors of the Nazi years and left the country shortly after the war ended. Proving himself an adept student, Schaller went on to earn a BS degree from the University of Alaska in 1955. Fascinated by wildlife, he enrolled in the zoology graduate program at the University of Wisconsin–Madison, working under the supervision of Professor John T. Emlen. Several factors, including concerns about possible extinction, led them also to choose the Mountain gorilla for study, and together they undertook a survey from March to the end of July 1959 to determine exactly where groups lived. In all, they demarcated sixty areas ranging in size from ten to 100 square miles each spread across 1,900 square miles in the eastern Congo, Rwanda, and Uganda. Taking a detailed census, however, proved to be impossible given time and labor constraints. As a best guess, they initially suggested a range of between 3,000 and 15,000 individuals, later downgraded to a more likely figure of 400 to 500.[24]

Plate 5.2. A Mountain gorilla group relaxing in the forest understory. ***Source:*** **George B. Schaller,** ***The Mountain Gorilla: Ecology and Behavior,*** **Plate 19 (University of Chicago Press, 1963, all rights reserved).**

Sponsorship by the New York Zoological Society (shortly to become the Wildlife Conservation Society) allowed Schaller and his wife to stay on for more than a year in order for him to study the ecology and population characteristics of gorillas on the Congo side of the Albert National Park, now called the Parc National des Virungas. The Rwanda portion has been renamed Parc National des Volcans. When possible, he hoped to make supplementary observations on other primates in the area, although, as it turned out, only the golden monkey *Cercopithecus kandti*, not to be confused with the golden monkey *Rhinopithecus roxellana* of China, lived at such high altitudes. In all, Schaller reported 314 visual contacts with gorillas during 466 hours observing them.[25] He tracked by himself, and his process of habituation, which means getting the animals used to a human's presence, involved making sure they could see him at a reasonably safe distance while at the same time trying not to make it obvious that he was watching them. Only in this way did Schaller believe that the collection of unbiased data would be possible. To avoid frightening his subjects and thereby making them wary of future contacts, he refrained from following groups when they departed from the original observation point. His book *The Mountain Gorilla: Ecology and Behavior* is loaded with information on the apes' characteristics, including their activities and behaviors within groups and as individuals. Seventy-four tables are appended, and the sum total of information provided primatologists with the most detailed look at free-ranging gorillas yet obtained. Unlike Osborn, Schaller did not mistake copulation for play and recorded several instances, and he became the first person to make detailed observations of what happened during intergroup contacts. None he saw ever resulted in "serious quarreling or fighting," and for the most part the gorillas appeared to be amiable.[26] Regarding conservation, Schaller concluded that while extinction did not appear to be "imminent, constant vigilance should be maintained by conservationists" to make sure that gorillas survived within the Virungas.[27] He did express special concern about the need to preserve their habitats, some of which were under threat from intrusions by farmers and cattle herders.

Like Donisthorpe, Schaller published a popular account, but unlike her effort, *The Year of the Gorilla* is substantial and attracted a large readership. In many ways, it's a travel book but one sticking to the facts. Freed from the strictures of writing for scientists, he could speak to his audience with the intent of drawing them into the lives of apes that he had come to respect and admire. For example, here's how he described his first meeting with a group:

> We sat watching each other. The large male, more than the others, held my attention. He rose repeatedly on his short, bowed legs to his full height, about six feet, whipped his arms up to beat a rapid tattoo on his bare chest, and sat down again. He was the most magnificent animal I had ever seen. His brow ridges overhung his eyes, and the crest on his crown resembled a hairy miter; his mouth when he roared was cavernous, and the large canine teeth were covered with black tartar. He lay on the slope, propped on his huge shaggy arms, and the muscles of his broad shoulders and silver back rippled. He gave the impression of dignity and restrained power, of absolute certainty in his majestic appearance.[28]

Schaller had no qualms about anthropomorphizing in order to bring gorillas closer to humans. The classic way of doing this, of course, is to provide the animals with names, just as we humans do with our pets. He chose to focus on characteristics and thereby convey something unique about each individual. Consequently, we meet the likes of Big Daddy, Splitnose, Mrs. Bad-eye, Callosity Jane, and Newcomer, among others. Certain behaviors could be rendered in human terms as well, so building a nest in which to sleep becomes "going to bed." Regarding an instance of play, he characterized one infant as being overcome with "*joie de vivre*."[29]

Most of the information about gorilla behavior in the book is conveyed in a chapter titled "A Gorilla Day." It's not an exciting read, for, as Schaller noted, "The gorilla's life consists of sleeping and feeding and sleeping some more. Only at irregular intervals does anything break the routine of its existence."[30] And most of these "irregular intervals" hardly constitute thrilling adventures.

Still, the graceful prose serves to convey both a sense of awe about and kinship with the gentle giants. Years later, he would remark, "No one who looks into a gorilla's eyes—intelligent, gentle, vulnerable—can remain unchanged, for the gap between ape and human vanishes; we know that the gorilla still lives within us. Do gorillas also recognize this ancient connection?"[31]

Surprisingly, given the alarm raised in *The Mountain Gorilla*, there's little in the book about conservation and the dire plight facing Mountain gorillas because of human intrusion into their habitats. The exception is Schaller's reaction to coming across cattle grazing in the vicinity of Kabara in August 1960. He described bringing his heavy walking stick down on the neck of one of the animals, causing its knees to buckle. Shortly thereafter, one of the guards with him set fire to a Batutsi hut and broke all the spears left behind by the fleeing occupants. At one point, Schaller threatened to shoot the cattle of a group of herders encountered within the park's boundaries. A few days later, guards in his presence killed about a dozen head and confiscated others. Justified actions? Certainly from the Batutsi involved, the answer would be "no," although the gorillas surely would say "yes" if they had their own voice.

Deteriorating conditions in the newly independent Congo brought Schaller's study to an end in September. He did return for ten days in August 1963 with a film crew from *Life* magazine to make a documentary, and by all appearances the park looked to have been well tended by the guards, with no cattle present. The gorillas, though, seemed shyer and more nervous than before. As he later put it, "They had obviously been harassed, their mountain peace shattered. Their idyll and mine was over."[32] Schaller would never study gorillas again. Instead, he turned his attention to an array of other animals, including African lions, tigers, snow leopards, jaguars, wild yaks, and giant pandas, a career track that has led him to become the world's most famous naturalist. Today, he's a senior conservationist with the Wildlife Conservation Society and vice president of Panthera, a global effort to save the last of the free-ranging big cats.

NYIRAMACIBILI

Nyiramacibili is the name used by some Rwandans for Dian Fossey. It's usually said to mean "the woman who lives alone in the forest," but there's no clear translation. The story of how she became Schaller's unlikely successor and most well-known and controversial gorilla researcher ever is still incomplete. Her book *Gorillas in the Mist* and the movie that followed starring Sigourney Weaver are devoid of details prior to the time she arrived in the Virunga region, and both were produced for popular audience appeal, with the movie handling facts rather casually. Two biographies appeared shortly after Fossey's death in 1985. *Woman in the Mists* by Farley Mowat is based largely on a collection of Fossey's surviving notes, diaries, and correspondence, now housed at McMaster University in Hamilton, Ontario. Other documents presumably are scattered among an array of private collections. Mowat presents a fairly detailed account of her life from the materials he could find, but his interpretation suffers from an overly admiring and sympathetic tone. Harold T. P. Hayes's *The Dark Romance of Dian Fossey* relies heavily on interviews. While more nuanced than Mowat's account, he focuses on the subject's mental state, and consequently gorillas take a backseat to her relationships with other people. A quite curious book is *Gorilla Dreams: The Legacy of Dian Fossey* by Georgianne Nienaber. It's basically a love story of author for subject. In the first half, Nienaber has Fossey speaking to gorilla favorite Digit in an afterlife as a way to relate her own story. The second half is written as an autobiography for the years 1978 to 1985, with Fossey still looking down from above. Although well researched and at times moving, Nienaber goes out of her way to defend her subject and place the blame on others for Fossey's failings and problems. She also takes potshots at those determined to have wronged Fossey, which includes just about everybody. In sum, a full-fledged, evenhanded biography of Nyiramacibili waits to be written. For purposes of this book, the details of Fossey's past aren't required: the basics will do. As far as her mind-set and actions are concerned, I need only deal with these as they relate to how she both impacted the study and lives of gorillas and brought them and people together.

The only child of a broken home, Dian Fossey seems to have loved all kinds of animals from an early age and eventually hoped to become a veterinarian when she enrolled at the University of California at Davis. The hard sciences proved to be Fossey's undoing during the second year of study, so she transferred to San Jose State to earn a degree in occupational therapy in 1954. At some point in time, Africa caught her attention, and she developed a strong desire to go there. Tired of waiting for the right opportunity to come along, Fossey assumed a huge debt and took a leave of absence from the Korsair Children's Hospital in Louisville, Kentucky, where she worked with some of the more disturbed cases. On September 23, 1963, she boarded a plane for an eagerly anticipated East African safari.

Eschewing a tourist-type adventure, Fossey hired a private guide named John Alexander to take her to the main game reserves. The actor William Holden, whom she met at his ritzy Mount Kenya Safari Club, recommended him. The longer the journey lasted, the more each grew to detest the other. In her book, she never refers to him by name; rather he's "The Great White" when disparagement is called for. Making the trip even more difficult, Fossey suffered from an array of health problems. Allergies and asthma plagued her, the latter exacerbated by chain smoking, and during a visit to Olduvai Gorge, she stumbled, turning an ankle badly in the process.

Shortly after the journey resumed, one of those seemingly small events that proved to be life changing occurred. While crossing the Serengeti, they ran into Dr. Jacques Verschuren, a biologist then working in the Congo, who recommended that she go see the Virunga gorillas. Fossey knew little or nothing about them, but a visit seemed too good an opportunity to pass up, and after heated debate she convinced Alexander to alter their plans and go there. Despite numerous obstacles, they reached the Kabara Meadow and, as good fortune would have it, met up with highly acclaimed naturalist photographers Alan and Joan Root. Charming when she had to be, Fossey convinced Alan to take her to where gorillas might be seen, and fortune again intervened when six came into view, an event that produced a vow by her to return and learn more about them.

At the end of the safari, Fossey was deeply in debt and therefore returned to work at Korsair, hoping to sell photos and articles of her East African adventures to raise extra money. As it turned out, the best outlet she could find was the *Louisville Courier Journal*, which carried three stories, including a Sunday supplement.

In early March 1966, Louis Leakey showed up in town to give a lecture on human origins. After its conclusion, Fossey got in line to say hello. Mary Leakey had tended to her ankle sprain at Olduvai Gorge, and it's unclear if she and Louis met at that time. Nonetheless, he agreed to a breakfast get-together in the morning and during the conversation about gorillas noted that they needed someone like Jane Goodall, currently researching chimpanzees at Gombe in western Tanzania, to undertake their study. So far, though, he had yet to find that someone. To Fossey's great surprise, several weeks later a letter from him arrived suggesting that she might, in fact, be the one he was looking for. Apparently, Leakey had been impressed by the pieces she had written for the *Courier Journal*, and he preferred female researchers for primates, thinking them more patient and less threatening than men. Fossey was thrilled and quit her job in anticipation of going back to Virunga country. But then anxiety set in as day after day went by without confirmation from Leakey. When it finally came, she learned that Leighton Wilkie, a benefactor who had helped underwrite Goodall's initial chimpanzee research, would contribute $3,000, just enough to begin a study. Leakey then pieced together small sums from several other sources, including the New York Zoological Society and the African Wildlife Leadership Foundation. He also submitted a proposal to the National Geographic Society (NGS), which soon became her primary sponsor. Additional funding came from the L. S. B. Leakey Foundation, which was formed following his death in 1972. Full of both excitement and uncertainty, Fossey left for the Congo on December 15, 1966. Before beginning her study of gorillas, she met with Jane Goodall and her husband, Hugo van Lawick, who took her on a tour of Gombe in order to get some ideas about how to carry out fieldwork, something Fossey had no experience in doing.

Plans called for spending two years at Kabara. That would be almost one year longer than Schaller's stay, and he hadn't learned everything, especially about the social life of gorillas. Confidently, Fossey vowed to "out-Schaller, Schaller." Things couldn't have started any better. Sanweke, who had served both Akeley and Schaller, signed on to be her teacher and tracker, and on January 9, 1967, they made contact with a group of nine gorillas. Discomforts and frustrations aside, all went reasonably well for the next several months, but these were dangerous times in the country, with Kivu Province harboring a secessionist movement that intensified when European mercenaries helped rebel leader Moise Tschombe take the cities of Kisangani and Bukavu. In the wake, all hell broke loose, as totally undisciplined troops killed, tortured, and raped at will. The deteriorating situation led President Joseph Mobuto to declare a state of emergency, and in a letter dated July 7, 1967, the director of the Parc advised Fossey to leave the area as quickly as possible.[33] Devastated, she had no recourse except to comply out of necessity, thinking that when things calmed down it would be possible to return. Basically ignorant of Africa other than its animals, Fossey had no idea of the political situation in which she had placed herself.

The next couple of weeks are murky. Fossey does seem to have been incarcerated in the town of Rumansabo, where later she would tell several people about having been abused and raped repeatedly by soldiers, although at first Fossey denied it, often casually remarking that "I think they were saving me for their major."[34] She said nothing about being raped upon reaching sanctuary at Travellers Rest after convincing her captors to take her there under a promise to pay them a substantial sum of money, which they never got, nor is there an account of this in *Gorillas in the Mist*.

Afterward, during a visit to Nairobi, Leakey tried to convince Fossey to study orangutans in Borneo or, perhaps, switch to Western gorillas. They had developed an intimate relationship by this time, with Leakey more the pursuer than the pursued. She, however, had become even more determined that her subjects would be Mountain gorillas. Both the Wilkie Brothers Foundation and the NGS agreed to a change of venue, which turned out to be within Rwanda's Parc National des Volcans at a place she named Karisoke after mounts Karisimbi and Visoke. The African Wildlife Leadership Foundation formed by Washington's Safari Club Conservation Committee in 1961 added further financial support. With the help of Belgian national Alyette de Munck, who ran a plantation in Rwanda, Fossey began setting up a base camp on September 24, 1967. At first merely three tents, it would be transformed into her home in Africa until her brutal murder there just after Christmas 1985.

Alyette de Munck became Fossey's close friend and served as a kind of mentor. She had spent many years in the Congo and knew most of the ropes about how to deal with people and conditions in that part of the world more generally. One could argue that without de Munck's friendship and help, *Gorillas in the Mist* would never have materialized. However, as often happened with Fossey, a falling out later occurred, and in this instance it produced a complete break in friendship.

A more ideal place to make contact with gorillas than Karisoke probably didn't exist, and by the end of 1968 Fossey had identified nine family groups. In 1972, she counted ninety-six gorillas in eight troops, one having broken up. It's common for troops to form and reform, depending largely on the competition and preferences of the silverbacks. The females can either stay together or seek acceptance by other troops. Fossey's first order of business involved habituation, and she soon discovered that the conventional strategy employed by Schaller of sitting still and watching had its limits as far as getting close to gorillas was concerned. Better, Fossey thought, to be more active and "elicit their confidence and curiosity by acting like a gorilla." She started by mimicking feeding and grooming practices and later added vocalizations that included the "deep belching noises" gorillas frequently make as they digest their highly fibrous foods.[35] Fossey claimed to have begun learning these while nursing two young gorillas (Pucker, a male, and Coco, a female) back to health. They had been captured for the zoo trade, and when her hopes for releasing them back into the wild came to naught, both were sent off to the Cologne Zoo, where they survived for nine years.[36]

Fossey's shining moment occurred when a young male, Peanuts, nestled up close and touched her hand, a first as far as anyone knows. The NGS photographer Robert Campbell was with her at the time, but the moment passed so quickly that it's not recorded on film. He did, however, manage to get a shot with the two of them lying close by one another, the event being featured in one of the articles she wrote for *National Geographic*.[37] Celebrity status quickly followed, and henceforth Fossey would find herself alternating between the remoteness of Karisoke and the fast lane of international speaking engagements and conference presentations. In addition, she needed to be away from her gorillas on occasion in order to secure financial support and complete a PhD at Cambridge University, which she received in May 1976. In 1979, Fossey accepted a visiting professorship at Cornell University in

Plate 5.3. Dian Fossey and Peanuts in their iconic moment together as captured on film by Robert Campbell. *Source:* Robert I. M. Campbell/National Geographic Stock.

Ithaca, New York, another boost to her academic credentials, always a sore spot in the competitive world of research. She spent much time in Ithaca between 1980 and 1983, using it as a base of operations for visits and lectures and to complete her book, which unfortunately didn't turn out to be quite the financial success she and the publisher had hoped for. Although Fossey did go on book tours, her heart really wasn't into them.

Habituation would eventually become a problem for her. For one thing, Fossey sought to make gorillas wary of Africans, claiming that they were the ones who killed them. This, of course, led to charges of racism. Fossey hardly helped her cause by frequently calling Africans "Wogs." In truth, she really didn't like them. Furthermore, she learned that close contacts with humans exposed gorillas to new diseases, such as hookworm, which could be lethal. Later it would be shown that habituation poses another threat, by leading groups to wander out of protected areas in search of cultivated food delicacies. Two unpleasant consequences have been documented when this happens: villagers respond by trying to kill the invaders, and gorillas sometimes retaliate, wounding or even occasionally killing villagers.[38]

Shortly after her arrival at Karisoke, Fossey began speaking out on the issues of gorilla survival and environmental preservation for all animals within the Parc. During 1968–1969, the protected area had been reduced from 500 square kilometers to 375 due to clearings to support the planting of pyrethrum. In the wake of bans on DDT and the search for organic pesticide alternatives, the plant's flowers had become a hot commodity, although this would last for only a short time due to the discovery of other synthetics less environmentally harmful than DDT. The removal of the land at lower elevations limited the area over which gorillas could range in search of food and made finding them easier. Although the Batwa, like other Rwandans, do not eat primates of any kind, they hunted an array of game for food and in the process sometimes ensnared gorillas in their wire-spring traps. In addition, more and more Batutsi had begun entering the forested area within the Parc in search of grazing lands for their cattle, further limiting the food sources available to gorillas. Their numbers were clearly in decline, down to perhaps 225, or more than half since Schaller's time.[39]

Administration of the Parc fell to the Office Rwandais du Tourisme et des Parcs Nationaux (ORTPN), and Fossey had managed to convince authorities to add more guards. But they were proving to be less than diligent, often just sitting around and occasionally taking bribes from the poachers and herders whom they managed to catch. This convinced her that Karisoke needed its own special police force to engage in "active conservation," which Fossey described as including "frequent patrols in wildlife areas to destroy poacher equipment and weapons, firm and prompt law enforcement, census counts in regions of breeding and ranging concentration, and strong safeguards for the limited habitat the animals occupy."[40] In this instance, the force would be largely herself, trusted locals, and young Americans and others who came to work and study at the Karisoke Center for Mountain Gorilla Research. She basically wanted to put fear into the hearts of potential transgressors and on occasion wore Halloween masks and acted the part of a witch to further frighten them. Such actions created ongoing tensions with ORTPN, and, of course, the Batwa and Batutsi resented her since they considered the resources of Parc part of their traditional rights. In the wider world, it led to accusations of Fossey acting the part of imperialist and taking the law into her own hands. She was accused of being a "gunslinger" and even murderer. While occasionally stating a desire to kill this or that poacher, no evidence has ever surfaced that she went this far. Fossey did, though, shoot a cow and on two occasions kidnapped children for short periods of time to make her point. Certainly with regard to the issue of park protection, she did "out-Schaller, Schaller."

Other rumors had Fossey being a manic depressive and an alcoholic. Although she drank heavily on occasion and suffered from bouts of serious depression, there's no clinical evidence to support either accusation. One thing can be said for certain: Fossey had a violent temper that at times became rage. Students, staff, and Rwandan officials all came in for their share of abuse.

Fossey's frustration with ORTPN intensified with the brutal killing of Digit on December 31, 1977, an act made worse by his head and hands having been cut off for sale. The perpetrators were never caught, and they may have done the deed out of revenge. Similar to Peanuts, Campbell had photographed a touching moment between Digit and Fossey. His death meant that now more than ever, gorilla preservation became the top priority, leading her to establish the Digit Fund to secure needed funds. She campaigned vigorously, and donations flowed in, especially after being featured by Walter Cronkite on his evening news broadcast. But Fossey soon found herself at odds with another organization, the Mountain Gorilla Project (MGP), an active supporter of tourism development, which Fossey bitterly opposed. A long, acrimonious dispute ensued not only about tourism but also about the MGP using Fossey and Digit for its own financial ends. More enemies thus surfaced, including several people she once trusted. As accusations and controversies intensified, many wanted her gone from Karisoke for good.

For all intents and purposes, by the late 1970s Fossey had abandoned research. The requirements of "active conservation" took up more and more time, and her health was in serious decline. Numerous broken bones, rotting teeth, hepatitis, sciatica of the hip, two abortions, and a serious kidney infection had all taken their toll. At one point, she almost died of rabies, getting the needed vaccinations just in the nick of time. Plagued as well by chronic emphysema and back pain, she simply lacked the strength and endurance required for serious fieldwork. Fossey did recover her health a bit during the time spent at Cornell, where demands were few, but not enough to take up the active life of studying gorillas when she returned to Karisoke in 1983.

Infirmities didn't kill her. Instead, early in the morning of December 27, 1985, she was found dead in her cabin, two panga blows to the head having done the job. Speculations about who had committed the murder filled the air during the ensuing days. Could it have been a poacher she had abused? Someone else she had offended? Rwandan officials blamed the American Wayne McGuire, the lone student in Karisoke at the time. Few believed this possible, but to err on the side of safety he left the country under advice as pressures for his arrest mounted. In absentia, a court found him guilty and imposed a sentence of death by firing squad. Fortunately for him, the United States and Rwanda did not have an extradition treaty. McGuire was eventually cleared, and Protais Zigiranyirazo, then prefect of Ruhengeri Province, went to jail on that charge along with crimes committed in the 1994 Rwandan genocide. The reason for killing Fossey may have been linked to the possibility that she was about to expose him for illegal gold trafficking. In 2009, the International Crimes Tribunal for Rwanda overturned his twenty-year sentence and ordered him acquitted of all charges based on "serious errors" during his trial for genocide.

Although Fossey wasn't a prolific writer, she did make important contributions to furthering knowledge about Mountain gorillas. Certainly no one before had had more intimate contact with them, and this allowed her to confirm their family-like social organization, feeding habits, and essentially peaceful nature. She, more than even Akeley and Schaller, helped dispel the notion of gorillas being savage beasts and showed that they could learn to tolerate human presence if not harassed. Beyond this, no one before had systematically documented gorilla vocalizations, and she provided the first detailed case studies of individual gorillas and their group dynamics.[41] Fossey also showed that males can be aggressive when it comes to protecting territory and families from other gorillas and in competing for dominance. She and other researchers documented nine cases of infanticide, all but one by silverbacks as they replaced or sought to replace a group leader.[42] Along with Alexander (Sandy) Harcourt, Fossey helped to expand knowledge about Mountain gorilla feeding habits, emphasizing the need to take into account differences in group age/sex compositions and other social factors.[43] She also coauthored a paper with Harcourt and Kelly Stewart showing that female Mountain gorillas changed groups, often more than once, whereas some young males emigrated from their groups to travel alone until such time as they attracted females.[44] The three also published an article on reproduction in the wild.[45] Toward the end of her life, Fossey helped to produce evidence regarding gorilla parasitology that eventually led to the formation of the Mountain Gorilla Veterinary Project of the Morris Animal Foundation.[46] It is still going strong today.

Dian Fossey's legacy, however, is associated more with having brought widespread public attention to Mountain gorillas and specifically to how ever-increasing encroachment by farmers, herders, and others could spell their doom. She did her best to humanize her subjects, notably by giving them endearing names that people could identify with, such as Peanuts, Beethoven, Effie, Uncle Bert, Macho, and Puck. Fossey also hit the lecture circuit to proselytize on behalf of Mountain gorillas and to collect money for support of Karisoke. She appeared in episodes of *Wild Kingdom* and as a guest on late-night television talk shows to plead her case. Then, too, it was Fossey's presence that drew photographers and film crews to Karisoke. These included naturalist David Attenborough for his *Life on Earth* series, which aired on television in 1979. He had come to do a segment on the opposable thumb, and while out looking for a simple shot of gorillas, he found himself face-to-face with an adult female. As recorded by him after the experience,

> There is more meaning and mutual understanding in exchanging a glance with a gorilla than with any other animal I know. Their sight, their hearing, their sense of smell are so similar to ours that they see the world in much the same way as we do. We live in the same sort of social groups with largely permanent family relationships. They walk around on the ground as we do, though they are immensely more powerful than we are. So if there were ever a possibility of escaping the human condition and living imaginatively in another creature's world, it must be with the gorilla. The male is an enormously powerful creature but he only uses his strength when he is protecting his family and it is very rare that there is violence within the group. So it seems really very unfair that man should have chosen the gorilla to symbolise everything that is aggressive and violent, when that is the one thing that the gorilla is not—and that we are.

In retrospect, I think it can be said that the photographic images spoke more loudly than words. Fossey used them to good effect in her talks and included many that she herself took in *Gorillas in the Mist*. But some of the most iconic are those shot by *National Geographic* photographer Robert Campbell, with whom she had an on-and-off-again relationship from April 1969 until March 1972. It might not be going too far to say that these images put Dian Fossey on track to become the public star she became, which, in turn, made Mountain gorillas stars.

No one filled Fossey's shoes after her death. And no one really had to in quite the same way because the case for saving Mountain gorillas had been made. Henceforth, efforts would be focused on furthering scientific understanding and developing more effective conservation practices.

THE MGP AND BEYOND

The MGP had its origins in the Mountain Gorilla Preservation Fund formed in Great Britain in early 1978 by John Burton and Alexander Harcourt, who filled in for Fossey at Karisoke during her 1980–1983 absence and, as noted, coauthored several papers with her. The Flora and Fauna Preservation Society (now Flora and Fauna International) provided initial backing, and later the African Wildlife Leadership Foundation (now the African Wildlife

Foundation) added its support along with the Worldwide Fund for Nature. Although the MGP closed down in 1989, having been absorbed into the broader International Gorilla Conservation Project, its activities have to be considered a success. During the first year of operation, more than 1,000 tourists paid to see Mountain gorillas, and the numbers kept growing, so much so that ORTPN began turning a profit, the income serving to cancel plans for a grazing scheme to take over one-third of the park area. In addition, the MGP made education within Rwanda a top priority, hoping to influence the country's leaders about the importance of saving gorillas and the general need for conservation. Up until this time, only the children of political and economic elites received more than a rudimentary education. The effort, along with those aimed at swaying public opinion, seems to have helped turn the tide in favor of preserving gorillas for future generations.

The success in Rwanda led Zaire, as the Congo had been renamed, to establish the MGP in its portion of the Virunga chain in 1984. Under the auspices of the Institut Zarois pour la Conservation de la Nature, with the support of the Frankfurt Zoological Society and the World Wildlife Fund, it adopted a "controlled tourism" policy that restricted visitors to a maximum six per day per one habituated gorilla family.[47]

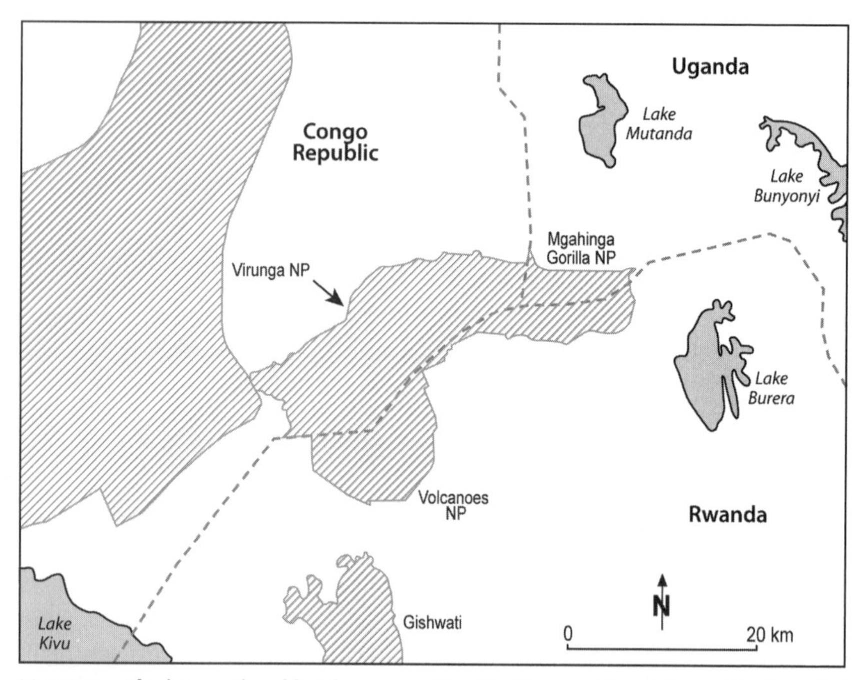

Map 5.1. The international borders separating the Virunga Mountain gorillas have led to very different outcomes regarding their study and human impacts.

The MGP's two most influential members on the ground were Bill Weber and Amy Vedder, both of whom had worked for Fossey and then fell out with her. She never published any serious accusations against them, but obviously confrontations took place, and Fossey certainly had words to say to others. To clear the air as both Weber and Vedder saw it, they launched a full-throttle attack on her in their 2001 book *In the Kingdom of Gorillas*, the primary reason that Georgianne Nienaber came to Fossey's defense.

Meanwhile, research and conservation work continued at Karisoke under the leadership of David Watts, who worked with Fossey in the late 1970s, and Dian Doren. Watts published a number of articles on gorilla behavior and reached out to form working relationships with the MGP and other projects in Zaire and Uganda. Financial support initially came from the Digit Fund, later named the Dian Fossey Gorilla Fund International. In 1991, Partners in Conservation teamed with the Fund. It helps provide money to pay the salaries of gorilla trackers and provides education to staff and families, including language training in French and English. Zoo Atlanta then subsequently became the new home for the Fund. Thus, as the 1990s got under way, the tide looked to have turned from the possible extinction of Mountain gorillas to their likely survival.

Human disasters suddenly turned optimism into pessimism. They started with fighting in Rwanda in August 1990, as Batutsi forces formed by exiles in Uganda under the banner of the Rwandan Patriotic Front (RPF) invaded to restore their leadership position usurped by a Bahutu coup after independence. Although defeated, some stayed, finding refuge in the Virunga region. They did not, however, bother the gorillas. Actually, orders had been given not to harm them in any way. Still, as a result of the turbulence, a George Schaller–led NGS film crew was forced to turn back after another round of fighting broke out between the RPF and government forces aided by French paratroopers. For a short while afterward, it looked as though things might cool down, but conflicts continued, leading to the slaughter of Batutsi in 1994, unleashed by the Bahutu *Interahamwe* ("those who fight together"). This prompted another RPF invasion, which succeeded in ousting the government. Although on a smaller scale, killings and other violence continued.

Under such conditions, gorilla research ceased. Karisoke itself was looted and destroyed in February 1993. Although it was rebuilt later that year, the staff decided to abandon Karisoke for good, following another destructive looting in 1996 by Bahutu rebels hiding in the Virunga forests. Remarkably, some Rwandan staff risked their lives to monitor the status of the gorilla groups, and only one killing seems to have occurred through all the turmoil. The Karisoke Research Center, though, continues. Relocated to Muhanzi (formerly Ruhengeri), it carries on Dian Fossey's legacy.

Things went badly for gorillas on the Congo side of the border as well. Starting in 1991, violence became the norm in the eastern part of the country, with eighteen gorillas reportedly killed in the late 1990s.[48] Yet despite the loss, Mountain gorilla numbers overall did not decline. A census in the late 1970s revealed that the Virunga gorilla population had stabilized at about 260 individuals after a precipitous fall in the 1960s due to civil war in the Congo that drove rebels and refugees into its portion of the Virunga region and, as noted, drove Fossey out.[49] Subsequent censuses and estimates show a consistent increase to the point where the number of Mountain gorillas in the four national parks is closing in on 900.

GORILLAS ELSEWHERE

Gorillas elsewhere have received only a fraction of the attention lavished on their Mountain cousins. The best known are the Eastern versions living in the Kahuzi-Biega National Park established in 1970 through efforts of the Belgian Adrien Deschryver, who somehow managed to maintain a semblance of order during the tumultuous 1960s and 1970s. Through his efforts, the park became the first one to establish organized tourism focused on gorillas. A singular moment boosting its fame occurred in 1975 when filmmaker Göta Dieter Plage photographed Deschryver trying to introduce an orphaned infant to the sights and sounds of the forest. Without warning, a gorilla group appeared, one he knew. Still, the silverback charged Deschryver, who dropped the infant, and the silverback scurried off with it, no damage having been done.[50] The event can be seen in Plage's movie *Gorilla*. Later, at an undocumented time and under suspicious circumstances, Deschryver died suddenly. A lingering suspicion is that someone poisoned him in an act of revenge.

In 1972, Alan Goodall (no relation to Jane) received permission from Deschryver to do research on the feeding and ranging behaviors of gorillas for his PhD Goodall originally had intended to work on Mountain gorillas under Fossey's supervision, and during the latter part of 1970 she trained him to study habituated bands and oversee conservation. Pleased with his performance, Fossey put Goodall in charge of Karisoke when she left in early 1971 for engagements in the United Kingdom and the United States. Upon her return in mid-March, things went sour between them because of, according to Goodall, some unspecified "misunderstandings." In all probability, these involved the killing of six gorillas under his watch. As a result, Fossey decided Goodall had to be let go. He wrote her several times hoping she would relent and allow him to finish his research, and eventually she did draft a letter of agreement for him to sign. Goodall, however, refused to do so because of conditions that barred his family from residing at Karisoke and a requirement giving Fossey control over any articles he might wish to publish. She could say yes, she could say no, or she could demand revisions.

With his previous research no longer of relevance, Goodall had to start from scratch at Kahuzi-Biega. He did manage to finish a study of the park's two habituated troops, focusing on their feeding and movement habits in one of his two major scientific publications. The other is a coauthored piece on Eastern gorilla ecology and conservation.[51] With reference to the latter, the authors saw habitat change due to human actions as the biggest concern. As we shall see, something else would soon become paramount.

Goodall also published a popularized account called *The Wandering Gorillas*. While engagingly written, the book contains little in the way of new knowledge about gorillas. And there's a rather strange claim about the film crew for *2001: A Space Odyssey* arriving in April 1972 to record "this pathetic side-shoot of the primate evolutionary line," namely gorillas. Not surprisingly, topography and rain hindered their filming efforts, although cameraman Helmut Barth supposedly did get almost an hour's worth of "good exposure," which included a charge by a silverback. *2001*, however, came out in 1968, with Barth not listed among the credits. Consequently, the film that the crew came to shoot for had to be a different one, and from what can be determined it never saw the light of day. As for Alan Goodall, Kahuzi-Biega was his last foray into the study of gorillas.

In 1980, UNESCO declared Kahuzi-Biega a World Heritage Site, and soon thereafter research began again under the leadership of Juichi Yamagiwa from the Kyoto University Primate Institute.[52] By then, the original two habituated groups had grown to five, and tourism was on the rise. Yamagiwa and his colleagues continued doing research, mostly again on diet, until disaster struck in 1994 with a huge inflow of refugees from the chaos in Rwanda. UNESCO responded in 1997 by classifying Kahuzi-Biega as a World Heritage Site in danger, but conditions worsened following the outbreak of civil war in the Democratic Republic of Congo in 1999, one result being the virtual elimination of four habituated gorilla troops.[53]

Although familiar from zoos across the world, the Western gorillas still residing in Africa remain largely undocumented, largely because they are so difficult to find due to forest density and widely dispersed populations. Usually, contact would come about suddenly, causing the startled gorillas to flee or attack, thus making study impossible. The best known until recently are those in Rio Muni, formerly part of Spanish Guinea, now Equatorial Guinea, due to some forty years of study by ethologist Jordi Sabatar Pi. In addition, he was responsible for bringing Snowflake, the world's only known white gorilla, to the Barcelona Zoo in 1964. We'll meet Snowflake in chapter 6.

In the mid-1990s, researchers discovered that food sources in open marshy areas called *bais* attracted visits by Western gorillas.[54] The opportunity to see animals without having to go through the problems of tracking them prompted Claudia Olejniczak to launch a study of the social life of troops congregating at Mbeli Bai in the Republic of the Congo's Nouabalé-Ndoki Reserve. Success has led other researchers to follow in her footsteps and study what have become habituated troops.[55]

Serious research finally began on the very hard to find Cross River gorillas in the late 1990s with K. L. McFarland's field study at the Afi Mountains Wildlife Sanctuary in Nigeria.[56] Unfortunately, one project came to a tragic end on June 29, 2010, when thieves killed primatologist Ymke Warren in the home she shared with her partner Aaron Nichols in Limbe, Cameroon. Currently, the Cameroonian Denis Ndeloh Etiendem, with support provided by the Wildlife Conservation Society and the Antwerp Zoo, is completing a study of the feeding ecology of groups within the Mawambi hills of southwestern Cameroon. An important outcome of his study will be more precise data on how human activities influence the eating habits of the Cross River gorillas.[57] If form holds, other researchers are bound to follow.

Since the 1950s, much has been learned about gorillas to help dispel long-held myths about them, and currently more research is ongoing that will add to what we know. Yet there's another way besides research and books many people have come to know gorillas. This is by seeing them in person or on film. Some became famous by attracting large audiences, and it's to them we now turn.

NOTES

1. A. G. Rhen, "Zoological Results of the George Vanderbilt African Expedition of 1934. Part I,—Introduction and Itinerary," *Proceedings of the Academy of Natural Sciences of Philadelphia* 88 (1936): 1–14.
2. Harold J. Coolidge Jr., "Zoological Results of the George Vanderbilt African Expedition of 1934. Part IV,—Notes on Four Gorillas from the Sanga River Region," *Proceedings of the Academy of Natural Sciences of Philadelphia* 88 (1936): 479–501.
3. Armand Denis, *On Safari: The Story of My Life* (London: Collins, 1963), 170–219.
4. Lucien Blancou, "The Lowland Gorilla," *Animal Kingdom* 58 (1955): 162–69.
5. Jeff Lyttle, *Gorillas in Our Midst: The Story of the Columbus Zoo Gorillas* (Columbus: Ohio State University Press, 1997), 11–32.
6. The story of Noell's Ark can be found in the little-known book by Anna Mae Noell, *Gorilla Show* (Tarpon Springs, FL: Noell's Ark Publisher, 1979).
7. Don Cousins, "Gorillas—A Survey," *Oryx* 14 (1978): 374–76.

8. Paul A. Zahl, "Face to Face with Gorillas in Central Africa," *National Geographic* 117 (1960): 132.
9. Walter Baumgartel, *Up among Mountain Gorillas* (New York: Hawthorne Books, 1976), 7–10.
10. Rosalie Osborn, "Observations on the Behavior of the Mountain Gorilla," *Primates* 10 (1963): 29–37.
11. M. Walter Baumgartel, "The Last British Gorillas," *Geographical Magazine* 32 (1959): 33–41.
12. Jill Wordsworth (Donisthorpe), *Gorilla Mountain* (London: Lutterworth Press, 1961), 11–12.
13. Jill Donisthorpe, "A Pilot Study on the Mountain Gorilla," *South African Journal of Science* 54 (1958): 195–217.
14. Jill Donisthorpe, "Gorilla," *African Life* 1 (1957): 38–41; "I Stalk Gorillas," *Personality* 20 (1958): 18–19.
15. Wordsworth, *Gorilla Mountain*, 11.
16. Oliver Milton, "The Last Stronghold of the Gorilla in East Africa," *Animal Kingdom* 60 (1957): 58–61.
17. Kinji Imanishi, "Gorilla: A Preliminary Survey in 1958," *Primates* 1 (1958): 73–78.
18. Masao Kawai and Hiraki Mizuhara, "An Ecological Study of the Wild Mountain Gorilla (*Gorilla gorilla beringei*)," *Primates* 2 (1958): 1–42.
19. Baumgartel, "The Last British Gorillas."
20. Ken Newman, "Saza Chief," *African Wildlife* 13 (1959): 137–42.
21. Zahl, "Face to Face with Gorillas in Central Africa," 117–18.
22. Charles R. S. Pitman, *A Game Warden among the Charges* (London: Nisbet & Co., 1931), 224.
23. Walter Baumgartel, "The Gorilla Killer," *Wild Life and Sport* 2 (1961): 14–17.
24. J. T. Emlin and George B. Schaller, "Distribution and Status of the Mountain Gorilla (*Gorilla gorilla beringei*)—1959," *Zoologica* 45 (1960): 41–52.
25. George B. Schaller, *The Mountain Gorilla: Ecology and Behavior* (Chicago: University of Chicago Press, 1963), 18.
26. Ibid., 119.
27. Ibid., 335.
28. George B. Schaller, *The Year of the Gorilla* (Chicago: University of Chicago Press, 1964), 34–35.
29. Ibid., 186.
30. Ibid., 197.
31. George B. Schaller, "Gentle Gorillas, Turbulent Times," *National Geographic* 188 (1995): 66.
32. Michael Nichols and George B. Schaller, *Gorilla Struggle for Survival in the Virungas* (New York: Aperture Foundation, 1989), 19.
33. A copy of the letter can be found in Farley Mowat, *Woman in the Mists: The Story of Dian Fossey and the Mountain Gorillas of Africa* (New York: Warner Books, 1987), 40.

34. Rosamond Halsey Carr, with Ann Howard Halsey, *Land of a Thousand Hills: My Life in Rwanda* (New York: Penguin Putnam, 2000), 154.

35. Dian Fossey and Robert M. Campbell, "Making Friends with Mountain Gorillas," *National Geographic* 137 (1970): 48–67.

36. For the story of Pucker and Coco, see Dian Fossey, *Gorillas in the Mist* (Boston: Houghton Mifflin Company, 1983), chap. 5.

37. Dian Fossey and Robert M. Campbell, "More Years with Mountain Gorillas," *National Geographic* 140 (1971): 574–85.

38. Thomas M. Butynski, "Africa's Great Apes," in *Great Apes and Humans: The Ethics of Coexistence*, ed. Benjamin Beck et al. (Washington, DC: Smithsonian Institution Press, 2001), 36–37.

39. Alexander H. Harcourt and Dian Fossey, "The Virunga Gorilla: Decline of an Island Population," *African Journal of Ecology* 19 (1981): 83–97.

40. Fossey, *Gorillas in the Mist*, 242.

41. Dian Fossey, "Vocalizations of the Mountain Gorilla (*Gorilla gorilla beringei*)," *Animal Behavior* 20 (1972): 36-53.

42. Dian Fossey, "The Imperiled Mountain Gorilla," *National Geographic* 159 (1981): 501–23, and "Infanticide in Mountain Gorillas (*Gorilla gorilla beringei*) with Comparative Notes on Chimpanzees," in *Infanticide: Comparative and Evolutionary Perspectives*, ed. Glenn Hausfater and Sarah Blaffer Hrdy (New York: Aldine Press, 1984), 217–35.

43. Dian Fossey and Alexander H. Harcourt, "Feeding Ecology of Free-Ranging Mountain Gorilla (*Gorilla gorilla beringei*)," in *Primate Ecology: Studies of Feeding and Ranging Behaviour in Lemurs, Monkeys, and Apes*, ed. Timothy Hugh Clutton-Brock (New York: Academic Press, 1977), 415–47.

44. Alexander H. Harcourt, Kelly Stewart, and Dian Fossey, "Male Emigration and Female Transfer in Wild Mountain Gorilla," *Nature* 263 (1976): 226–27.

45. Ibid. See Also Alexander H. Harcourt, Kelly Stewart, and Dian Fossey, "Gorilla Reproduction in the Wild," in *Reproductive Biology of the Great Apes*, ed. Charles E. Graham (New York: Academic Press, 1981), 265–79.

46. Harcourt and Fossey, "The Virunga Gorilla," 83–97.

47. Conrad Aveling and Rosalind Aveling, "Gorilla Conservation in Zaire," *Oryx* 23 (1989): 64–70.

48. Andrew J. Plumptre and Elizabeth A. Williamson, "Conservation-Oriented Research in the Virunga Region," in *Mountain Gorillas: Three Decades of Research at Karisoke*, ed. Martha M. Robbins, Pascale Sicotte, and Kelly J. Stewart (Cambridge: Cambridge University Press, 2001), 381.

49. Bill Weber and Amy Vedder, *In the Kingdom of Gorillas: Fragile Species in a Dangerous Land* (New York: Simon & Schuster, 2001), 101.

50. The video can be seen at http://www.youtube.com/watch?v=MKkZawWUqTQ.

51. Goodall, "Feeding and Ranging Behavior of a Mountain Gorilla Group (*Gorilla gorilla beringei*) in the Tshibinda-Kahuzi Region, Zaîre," in Clutton-Brock, *Primate Ecology*, 449–79; Alan G. Goodall and Colin P. Groves, "The Conservation of Eastern Gorillas," in *Primate Conservation*, ed. His Serene Highness Prince Rainier III of Monaco and Geoffrey H. Bourne (New York: Academic Press, 1977), 599–637.

52. Juishi Yamagiwa and N. Mwanza, "Day-Journey Length and Daily Diet of Solitary Male Gorillas in Lowland and Highland Habitats," *International Journal of Primatology* 15 (1994): 207–24; Juishi Yamagiwa et al., "Seasonal Change in the Composition of the Diet of Eastern Lowland Gorillas," *Primates* 35 (1994): 1–14.

53. Juishi Yamagiwa and John Kahekwa, "Dispersal Patterns, Group Structure, and Reproductive Parameters of Eastern Lowland Gorillas at Kahuzi in the Absence of Infanticide," in Robbins, Sicotte, and Stewart, *Mountain Gorillas*, 116–17.

54. Richard Parnell, "Forest Clearings: A Window into the World of Gorillas," in *World Atlas of Great Apes and Their Conservation*, ed. Julian Caldecott and Lera Miles (Berkeley: University of California Press, 2005), 113–14.

55. Emma J. Stokes, Richard Parnell, and Claudia Olejniczak, "Female Dispersal and Reproductive Success in Wild Western Lowland Gorillas (*Gorilla gorilla gorilla*)," *Behavioral Ecology and Sociobiology* 54 (2003): 329–39; Thomas Breuer et al., "Physical Maturation, Life-History Classes and Age-Estimates of Free-Ranging Western Gorillas—Insights from Mbeli Bai, Republic of Congo," *Journal of Primatology* 71 (2009): 106–19.

56. K. L. McFarland, "Ecology of Cross River Gorillas (*Gorilla gorilla diehli*) on Afi Mountain, Cross River State, Nigeria," PhD diss., City University of New York, 2007.

57. Denis Ndeloh Etiendem, personal communication.

Chapter Six

Famous Gorillas

ON DISPLAY

Famous gorillas come in two types, real ones and fictional counterparts, most notably those on the silver screen. Digit is the only gorilla in the wild to have attained worldwide stardom, and in the early days, as we've seen, those put on display, either privately or in zoological gardens and circuses, didn't live long enough to develop large followings. This started to change with John Daniel (Johnny Gorilla to some). As the story goes, in 1916 hunters in Gabon came upon two males, a large adult and a juvenile.[1] When attempts to capture both failed, the adult was shot and killed, the juvenile then easily taken. With World War I in progress, a sea voyage to Europe couldn't be risked, so he wound up being kept at a small coastal encampment along with several German prisoners. When the war ended, the youngster was shipped to London, his owners hoping that a zoo would purchase him. No takers could be found due to cost and the likelihood of early death. Eventually, the department store of Derry and Toms in Kensington bought him to exhibit in a window, but he soon became despondent and stopped eating. Retired army Major Ruppert Penny took pity on the poor creature and gave the store the 300 pounds sterling it had paid the original owners. Penny handed him over to his niece Alyse Cunningham to nurse back to health and socialize at their home in Uley, Gloucestershire.

Cunningham named him John Daniel and converted a large room into a cagelike facility. He hated being in it at night, so she allowed him to sleep on a bed in an adjacent room. During much of the day, John Daniel had the run of the house, and he learned to eat sitting at a table. Visitors came to see him, and John Daniel especially enjoyed being in the company of children, often walking hand-in-hand with them. Toys and chickens also provided fascination, and he liked sitting by the fireplace to watch the flickering flames. According to one visitor,

> When I had the pleasure of making John's acquaintance in 1920, he was sleeping on a camp bed in Penny's room beneath blankets that he put over or took off himself just the same as you and I. He was scrupulously clean in his habits, and acted in this respect in the same manner as other members of Major Penney's household. He could unhasp and open the window, open the door or shut it when told to, and put on the electric light. He could drink out of teacups and put them back carefully on the tray, and many other intelligent things besides.[2]

From time to time, Cunningham took John Daniel by taxi to show at zoos and other public places, where he always drew large crowds.

Plate 6.1. John Daniel as displayed at the American Museum of Natural History. *Source:* American Museum of Natural History, image 39409, front view.

By the close of 1920, Cunningham knew that John Daniel had become too big to keep in the house. Reluctantly, in early 1921 she sold him to animal dealer John Benson for a thousand guineas under the pretext that his new home would be a Florida animal park. Instead, Ringling Brothers bought John, labeling him the "civilized gorilla." But John Daniel disliked circus life, usually ignoring visitors who came to see him. Soon his health began to decline, and on April 17, 1921, in a room specially arranged for him at Madison Square Garden, John Daniel died, joining the six other gorillas that had succumbed shortly after reaching the United States. Cunningham had been sent for in the hope that her presence would prove useful, but she didn't reach New York City in time. Many attributed his passing to a broken heart, although the final verdict pinpointed pneumonia. That he was fed a meat-rich diet probably didn't help his disposition or health. After being examined by a host of surgeons and other medical personnel, John Daniel wound up as a mounted gorilla exhibited at the American Museum of Natural History.

Cunningham tried to keep his name alive in public with a John Daniel II, called Sultan when she bought him in 1923. He toured successfully for a while but never achieved the fame of the original before dying in 1927.

A bevy of reporters and photographers greeted Ben Burbridge's Miss Congo, the "Miss" soon being dropped from her name, upon arrival at a New York City pier on October 15, 1925. After all, she was presumed to be America's first Mountain gorilla and the only captive female gorilla worldwide at the time. But according to an observer, Congo merited hardly a second glance riding in a taxi for a short stay at the Central Park Zoo.[3] From there, Burbridge took Congo by train to Shady Nook, his brother Jim's bucolic home beside the St. James River in Jacksonville, Florida. Jim's wife Juanita looked after her most of the time. His various experiences with gorillas convinced Ben that they should not be "in intimate association with humans," as this would expose them to diseases like colds and influenza. "Fresh air, exercise, and proper diet" were required, and they should not be put in solitary confinement. He felt that a young chimpanzee would make a good companion.[4] Instead, while living at Shady Nook, Congo had dogs to play with outside her cage. On occasion, she got a bit too rough for their liking.

Not long after arriving at Shady Nook, psychologist and primate specialist Robert Yerkes entered Congo's life. He learned about her on Thanksgiving Day 1925 when Carl Akeley came to visit him at the Yale University Primate Laboratory. Yerkes longed to study gorillas as he had other apes and in eager anticipation left by train for Florida on December 10. The Burbridges were only too happy to have Congo be his subject, so Yerkes immediately began his psychological experiments, "devised primarily to display the nature and degree of Congo's adaptivity, to reveal whatever measures of insight or understanding she might possess, and to elicit characteristic affective expressions and adjustments."[5] He did so for six weeks on this occasion and came back from mid-January to mid-March of the next year to administer follow-up testing. Congo was quite cooperative, except when it came to taking anthropomorphic measurements, which she vigorously resisted.

By this time, Congo had become too big and strong to keep at Shady Nook, so at the end of March, John Ringling North purchased her. For several months, she lived in a large cage on the grounds of Ca'd'Zan, the Gilded Age mansion of John and his wife Mable in Sarasota. The rest of her life she spent at the nearby winter quarters of the circus, surrounded by other animals, both wild and domestic, and with strange people almost constantly in view. Yerkes returned at the beginning of 1928 for further studies of Congo and found that she had adapted well, never showing signs of fear or anger in spite of all of the commotion. Toward the end of this session, she lost her appetite and died on April 23 from what was determined to be acidosis resulting from colitis, the blame being put on an excessive carbohydrate diet heavily based on fruits. Congo loved them and disliked vegetables. She lives on, though, as the most thoroughly studied gorilla of all time.

Readers interested in the details of Yerkes's methods and findings can find them in his three-part series *The Mind of a Gorilla*. Here, I'll just include a comment he made about Congo's general mood and how it differed from the opinion then held by many people about gorillas. "The terms sullen and morose often applied to the adult of the species," he noted, "certainly are not applicable to Congo, for she is habitually placid, self-dependent and reasonably superior to the accidents of her captive environment."[6]

Bamboo became the first zoo gorilla to gain a substantial following. One reason is that he lived to be over thirty years old, with the vast majority of his life spent at the Philadelphia Zoo. Brought there in August 1927 weighing just eleven pounds, Bamboo thrived under good care that allowed him to surpass by a wide margin the ten years in captivity achieved by reigning champion Bobby at the Berlin Zoo. No zoo gorilla in the United States until him had survived for more than a year. Also, during his early days at the Philadelphia Zoo, Bamboo had a friend, a female chimpanzee called Lizzie, to provide companionship and stimulation.[7] Unlike Bobby, who possessed a mean streak, Bamboo was outgoing and loved to play, which, among other things, involved throwing tomatoes and watermelons at visitors with an underhand-like pitch. One day, he is reputed to have splashed five photographers with a well-aimed melon.[8]

Bushman's celebrity status surpassed even Bamboo's. He had been captured as an infant in French Cameroon in 1929 by animal collector Julius L. Buck, not to be confused with the other famous collector and circus star Frank "Bring 'em Back Alive" Buck, and turned over to an American missionary couple in Yaoundé who found a local woman to nurse him until he was judged strong enough to endure an Atlantic crossing. Bushman was destined to join the large menagerie, including numerous chimpanzees and orangutans, kept by Madame Rosalía Abreu at her sumptuous estate Quinta Palatina on the outskirts of Havana, Cuba. In fact, the place resembled a private zoological garden more than it did a menagerie.[9]

But Buck and Abreu had a falling out over price, and Bushman eventually wound up at Chicago's Lincoln Park Zoo in 1930. From the outset, he displayed qualities of gentleness and playfulness, an automobile tire hanging from the top of his cage being a favorite toy.[10] Thousands lined up to see him virtually every day, and he seemed to relish performing for them, his antics making regular copy for newspapers and magazines. So famous did Bushman become that in 1946 he was judged to be the most valuable animal in captivity.

Bushman grew to be huge, reaching nearly six feet in height and topping out in his prime at around 550 pounds.[11] When news surfaced about his failing health, the crowds grew even larger, a reported 120,000 visitors coming by on a single June day in 1950 to wish him well. Bushman did get better for a while, but he died on New Year's Day 1951 after suffering from a heart condition, arthritis, and a vitamin deficiency that caused him to lose weight in his last couple of years. For weeks afterward, people walked by his cage saying their fond farewells to a picture of him set inside. Bushman, though, is still with us, as his skin was preserved and mounted for display in Chicago's Field Museum. Mysteriously, his skeleton disappeared, never to reappear.

Susie deserves mention if only because of having been the first gorilla to fly across the Atlantic Ocean. At about one year of age, she was captured in 1927 in the vicinity of Lake Kivu, thus making her of Eastern gorilla descent.[12] Transferred to the Atlantic coast for shipment to Europe and also initially called Congo, she wound up being put on display at several sites along the French Riviera, where her name became Susi. A German wild animal dealer in Hannover bought Susi to add to his collection and a little later sold her to a New Jersey dealer who arranged for her to cross the ocean aboard the *Graf Zeppelin*, which reached the company's Lakehurst depot on August 4, 1929. Following three days of being displayed at an unnamed New York City department store, she toured much of the United States and parts of Canada under the slightly altered name of Susie. When the touring days came to an end in 1931, the Cincinnati Zoo purchased her, and she spent the rest of her life there as a major attraction. Taught to use silverware and napkins, Susie ate regularly with her trainer William Dressman, learned how to brush her teeth, and liked to play catch. Gentle to the core, Paramount News featured her in "A Day in Susie's Life." Unlike other gorillas to date, she allowed Dressman to take detailed growth measurements. Susie eventually reached sixty-two inches in height and 458 pounds in weight, making her the largest female gorilla yet known.

After Susie suddenly died in 1947 from what proved to be leptospirosis, the Cincinnati Zoo donated her body to the University of Cincinnati, where she served as a specimen in introductory biology classes. In 1974, a fire broke out in the laboratory storeroom, and Susie's remains went up in smoke. So she's now just a memory, a pleasant one to be sure for all who knew her.

In early 1931, Arthur Phillips, captain of the *West Key Bar* plying the Atlantic trade between Africa, Europe, and the Americas, acquired a sick, orphaned baby gorilla near the mouth of the Congo River. His mother had been killed while raiding a plantation, and the villagers took him in. Phillips brought the youngster, who had been given the name Massa, pidgin for "master," to Gertrude Lintz (née Davies), wife of William Lintz, a Brooklyn doctor of internal medicine, hoping she would be able to nurse him back to health. An eccentric animal lover, Mrs. Lintz had gained a portion of fame by raising champion St. Bernards and later accumulated a menagerie, including carrier pigeons, rabbits, a great horned owl, a leopard, and chimpanzees that she had purchased from Phillips earlier on. Without hesitation, she agreed to give it a try, later saying that the earlier sight of a circus gorilla had convinced her that she "was one woman who could love a gorilla enough to keep it well."[13] With her devoted care, Massa eventually recovered from what turned out to have been pneumonia.

Later in 1931, Phillips arrived in New York harbor with another orphan who had been living with a missionary couple for a year. A disgruntled seaman took revenge on Phillips by spraying the youngster in the face with nitric acid with the intent of killing him. He survived the attack but was clearly in need of medical attention, and therefore, once again, Phillips contacted Gertrude Lintz. Upon reaching the ship, she found him "cowering in a corner," nearly blind and in serious pain. A quick trip to the Lintz home on Shore Road followed, where her husband called in an eye specialist who, after doing a thorough examination, said that drops would restore his vision to normal in due course. A plastic surgeon then helped repair damage to the face, the one permanent scar being a curl on the left side of the upper lip, which gave him a permanent snarl that would prove to be part of his later fame. Gertrude Lintz named him Budha, soon to be replaced by Buddy.

The two gorillas thrived, being raised like children. They wore shoes, sat in high chairs, and, like John Daniel, ate at a table. Mrs. Lintz had grown particularly fond of Massa, noting how affectionate he was and always "full of fun and charming, playful ways."[14] When he was at one point crippled by an infection, she designed a form of physical therapy that restored Massa's limbs to full strength. But a turning point came in 1935, when she inadvertently surprised Massa, who was cleaning the kitchen floor, a chore she had taught him to do. He turned quickly and struck out, not knowing who was there. The deep wounds required seventy stitches to close. As a result, Massa joined Bamboo at the Philadelphia Zoo, where officials thought the latter needed a companion. But fears soon developed about the two males possibly harming one another, so each was placed in a separate cage. Massa loved devouring birthday cakes, and he eventually outdid Bamboo by living to the ripe old age of fifty-four, showing that the Philadelphia Zoo had hit upon the secret of how to care for gorillas properly. A closer look at its history of keeping gorillas will be provided in chapter 7.

Two years after Massa left the Lintz household, Buddy's time to go arrived. Although he never hurt anyone, several episodes, including escapes from his cage, one of which involved jumping into bed with Mrs. Lintz during a storm, made it clear that he couldn't be kept safely at home anymore. On December 4, 1937, she wrote to John Ringling North, offering Buddy and two chimpanzees to the circus for $10,000. John and his brother Henry went to have a look, the latter remarking, "Behind the bars glowered the most frightful face I have ever looked upon. A tremendous hairy head, great dripping fangs, and horrible sinister leer of the acid-twisted mouth." When Buddy shook the cage and grabbed out at Mrs. Lintz, who, Henry said, "ran for her life," John became convinced that the circus needed to have this particular gorilla.[15] It hadn't had one since Congo's death in 1928. The name Buddy, however, didn't match their plans to draw huge crowds hoping to see a ferocious beast, so they chose Gargantua, and later added "The Great."

As an interesting aside, Gertrude Lintz claimed to have figured out the meanings of "drum codes" or chest beating from Buddy:

> Contentment: The full-belly beat is done on the happy stomach with the flat of the hand in slow, alternate slaps. The eyes roll from side to side, and usually there is a small smile on the lips. My gorillas used this as a general sign of lazy well-being.
>
> Intense joy: Excited, rapid beats on the chest with both fists at once, in increasing tempo.
>
> Rage: A loud tattoo on the lower ribs with both fists tightly clenched and striking together so fast that there is hardly a pause between beats. This is the war drum and should not be heard in the United States. It is the adult male's warning to the clan to scatter, and to the enemy to look out, and it can be heard a mile.[16]

It helps to have a good imagination.

Gargantua's public debut came in April 1938, and both *Life* and *Newsweek* put him on their covers. He was playful at first—tossing and catching a softball and engaging in tugs-of-war. But eventually Gargantua began to live up to his name. On one occasion, he grabbed his trainer, who, with help, managed to escape, although nearly in an unconscious state. At another time, he grabbed and began biting John Ringling North's arm and let go only after being beaten on the head.[17] His view of life was hardly helped when a discharged employee tried to poison him with disinfectant sweetened by chocolate syrup. The concoction severely burned his stomach linings, and as result he lost eighty pounds. Many weeks of care were required to restore him to health.

As Gargantua grew larger and stronger, it became clear that the original cage was not sturdy enough to hold him, nor did it have suitable temperature control. North contacted the Carrier Corporation in Syracuse, New York, to build a special wagon cage. When completed, it measured twenty-six and a half feet in length and contained two compartments so that one could be cleaned. Thick glass panels outside covered steel bars, and thermostatic controls maintained a temperature of seventy-six degrees Fahrenheit and kept the humidity at 50 percent. It would serve as Gargantua's hermetically sealed home for the rest of his life, affording protection from human infections and protecting the public from him.

With a massive publicity campaign that billed Gargantua as "The Largest Gorilla Ever Exhibited" and "The World's Most Terrifying Living Creature!,"

Plate 6.2. Poster by the Ringling Brothers and Barnum & Bailey Combined Shows designed to lure customers into paying to see Gargantua. *Source:* Photofest.

he became the circus's biggest attraction and largest revenue earner as it crisscrossed the continent during the eight-month-long show season. His popularity probably saved the company from bankruptcy. Although Gargantua looked ferocious and on occasion acted the part, most of the time he didn't do much of anything except play with his tire toy and watch people as they watched him.

Circus management thought that he needed a female both as a companion and to produce a gorilla family. They found one in Havana that Maria Hoyt had adopted in 1932 while on a safari with her husband Kenneth in the French Congo. One of their goals was to obtain a large gorilla specimen for the American Museum of Natural History, and when they came upon a family, the professional hunter with them killed the silverback. Villagers then netted the rest of the family and speared eight to death, leaving a lone two-month-old female survivor that Maria took into her arms and named M'toto (Kiswahili for "young child" but without thc apostrophe; later, more commonly, she became Toto). The Hoyts then turned the infant over to a local woman to be nursed and cared for. Impressed by Toto's gentle disposition, the Hoyts decided to bring her to Paris with them. During their stay at a hotel on the Rue de Rivoli, she came down with pneumonia. They called in a pulmonary specialist who successfully treated her. After the crisis had passed, the Hoyts took Toto to the seashore near Bordeaux to recuperate, her presence causing a sensation among the locals seeking to have a look. The Hoyts then made a decision to relocate in Havana instead of returning to their Connecticut home. Both liked the city and considered its warmer climate better suited for the youngster, which they now wanted to keep. Madame Abreu's successes with chimpanzees and orangutans gave them confidence that all would be well.

Similar to what happened with Buddy, the Hoyts raised Toto like a child. She had her own separate house that contained a playroom measuring fifteen by twenty-five feet and a bedroom ten by fifteen feet. A forty- by eighty-foot iron-barred enclosure adjoined it. According to Maria's daughter,

> Toto loved my mother. Whenever she went out, even for a few hours, the baby would run joyfully to meet her and would put her arms around her and give her a smacking kiss just as she did to me. She would sit for long periods gently stroking mother's white hair which seemed to fascinate her and never once pulled or deranged it. Mother taught her how to make pictures in the air with her finger, a circle for a face, three dots inside it for eyes and a nose and slash for a mouth. Often Toto would greet her by making this diagram in the air with an extended forefinger as though it were a sign between them of joint membership in some secret organization.[18]

She also adored a pet cat. Female gorillas, in particular, often have an affinity for small animals.

Plate 6.3. Toto ever so gently holding her pet cat. ***Source:*** **Hoyt,** ***Toto and I.***

Visitors flocked to see the only gorilla in Cuba, but the usual problem surfaced: Toto's growing size and strength. One day, in fact, she accidentally broke both of Maria Hoyt's wrists. Her husband Kenneth died in June 1938, and when Toto's handler became seriously ill, Maria was left to take care of matters pretty much by herself. She couldn't, however, manage to do so, and Toto began escaping from her quarters on a regular basis. Still, she persisted with Toto, turning down many offers from zoos and circuses. But when one escape nearly led to a tragedy, she knew the time had come for the young gorilla to go. The question was, where? After due consideration, Maria accepted an offer from John Ringling North as the best of several alternatives.

Toto arrived in Florida during February 1941 and became known as Mrs. Gargantua, following a wedding ceremony held by the circus on Washington's Birthday. Maria served as matron of honor, Schrafft's provided the cake, and the processional from Wagner's *Lohengrin* filled the air, with numerous reporters scurrying about to record everything.[19] In point of fact, the two gorillas never did share a cage. Instead, they occupied adjacent ones. Right from the start, neither liked the other, and fears developed about Gargantua harming Toto if placed together. Thus ended the idea of the circus being able to display the first gorilla family in captivity. Still, it carried on with the charade of marriage, sending out Christmas greetings showing a happy couple and using them to promote the sale of war bonds.[20]

As the years went by, Gargantua's health began to decline, and so too did his temperament, to such an extent that no one could go near to examine and treat him. By June 1949, he looked to be in particularly bad shape, and on November 25, 1949, workers found him dead while the circus was in Miami. The afternoon edition of the *Miami Daily News* carried the headline "Gargantua Dies," and other newspapers made the event a front-page story. An autopsy performed at Johns Hopkins University Hospital revealed double pneumonia, plus a range of other disorders, including a kidney infection and four rotted wisdom teeth.[21] Gargantua's skeleton was sent to Yale University's Peabody Museum for reconstruction and exhibition. Toto carried on, dying peacefully of old age on July 17, 1968. Shortly thereafter, the Ringling Brothers and Barnum & Bailey Combined Shows, which still tours today under the banner "The Greatest Show on Earth," added two baby gorillas, Mademoiselle Toto and Gargantua II. Neither, though, achieved the fame of the originals. Circus audiences had declined in the face of competition from other forms of entertainment, and the shows resorted to using fake gorillas among their acts, if they wished to exhibit them at all. The lesson had been learned—real gorillas are not natural-born performers like chimpanzees, and they cost a lot more to purchase and maintain.

An attempt to resurrect Buddy can be seen in a 1997 movie of the same name. It's a silly production and laden with errors from the beginning, when Gertrude Lintz, played by Rene Russo, rescues an infant named Buddy from a zoo, to its end, when he's shown romping off to be with others of his kind in an idealized gorilla heaven. Nothing is said of Gargantua, and viewers are told that Buddy lived to be fifty-four years of age, confusing him, obviously, with Massa. In fact, he's a collage of the two in the movie. As you can imagine, Buddy isn't portrayed by a real gorilla, and he's shown doing things like walking upright most of the time and serving appetizers to guests. This is a movie to avoid if you want the truth about gorillas and Buddy in particular.

Congo, renamed Mbongo after Alumbongo (the place where he was captured), and Ingagi, the two young gorillas that Martin and Osa Johnson sent to the San Diego Zoo, need to be singled out for attention. The zoo's executive secretary at the time was Belle Benchley. She had been hired in 1925 as a temporary bookkeeper and over the next several years worked her way up the hierarchy to become its director from 1941 to 1953, the first woman to head a zoo.[22] In the process, Benchley helped turn a once small, obscure facility into a world-class one. Mbongo and Ingagi certainly helped the cause. Both were older when they arrived than most zoo gorillas when they died, Mbongo being about five years of age and Ingagi closer to six. As noted in chapter 4, they were Eastern gorillas, not, as advertised, Mountain ones. Benchley took them under her wing, spending much time in their presence, which included a daily offering of grapes, corn, dried prunes, raisins, and special treats. A huge, specially constructed cage included tree stumps for the gorillas to sit upon and ropes for swinging, and they had hay from which to make nests. Another decision Benchley made was to minimize human contact. Neither of them received special handling or training, not out of fear but rather because she wanted Mbongo and Ingagi to be as gorilla-like as possible. Benchley also turned down a lucrative offer to sell them to a circus.

Naturally enough, scientists came to take measurements and make observations. C. R. Carpenter from Columbia University did so in July and August 1934 and discovered that Mbongo, the smaller of the two, was a male and not a female as the Johnsons and even Benchley thought. He also noted their playfulness with themselves, objects, and each other and concluded that the two were well adapted, pointing out that "references in the literature which describe the gorilla as being sullen, ferocious, aggressive and melancholy are out of line with the temperament of this study, and I am inclined to think such descriptions may be artifacts of abnormal conditions of the subjects studies and inaccurate descriptions."[23]

On most days, crowds gathered around the cage. Ingagi was shy and a bit retiring but not so Mbongo, who loved to perform, especially before children, and enjoyed exploring new things introduced into the cage. Those looking on couldn't help but notice what Carpenter had. Their general good natures led Edgar Rice Burroughs and Johnny Weissmuller to study them for the Tarzan books and films. Charles Gemora, known for his portrayals of Hollywood gorillas, also came to observe Ingagi and Mbongo in an attempt to get his performances closer to reality.

On the unfortunate side of the equation, both gorillas continued to gain weight, despite being fed mostly fruits and vegetables, supplemented by milk and eggs, which they liked. Neither touched meat when offered. At his death on March 15, 1942, Mbongo weighed 618 pounds, the proximate cause shown to be the fungus *Coccidiosis immitis*, likely introduced by a zoo visitor. Ingagi tipped the scales at 636 pounds shortly before succumbing on

January 12, 1944, from a coronary condition. Belle Benchley stepped down from her post in 1953 but kept active in zoo and animal affairs until passing away on December 19, 1973. Carved on her headstone is the image of a smiling gorilla. The zoo, however, has continued to grow in both size and status and today consists of two parts: the zoo proper in Balboa Park and the larger Safari Park in Escondido. All told, they contain more than 3,700 animals, among which are eleven gorillas in two troops.

The honor of being the first gorilla ever born in a zoo goes to Colo.[24] Her birth at the Columbus Zoo, however, wasn't a planned one. Indeed, the zoo's two adult gorillas, Baron Macombo and Millie Christina, supplied by "Gorilla Bill," were kept apart from one another because of several incidents when he displayed aggressive behavior in her presence, thus seeming to confirm the prevailing view that gorillas should be separated for their own good. Warren Thomas, a veterinary student working as a part-time keeper at the zoo, felt that the two might get along after noticing them acting playfully at the bars of their adjacent cages. Without asking permission, he convinced another keeper to help him put Macombo and Millie together at night and then separate them early the next morning before others could notice. Sure enough, they mated, and shortly thereafter Millie showed signs of being pregnant. No one knew anything at the time about gorilla births, and by lucky accident on the morning of December 22, 1956, Thomas found a newborn still encased in its amniotic sac lying on the floor of Millie's cage. She demonstrated no interest in her progeny, and Thomas had little trouble moving the mother to an adjacent cage. He then removed the sac and began mouth-to-mouth resuscitation to get the baby breathing properly. After that happened, Thomas put her in a cardboard box on top of clean rags and took it to the warmest spot at the zoo: the boiler room. Infant formula recommended by a pediatrician provided her first food, and the next day she was placed in an incubator and from there went into a specially constructed nursery with around-the-clock care for the next six months.

The event catapulted the modest Columbus Zoo and Colo, the winning name in a competition that drew some 7,500 entries, to star status. Zoos from around the world began asking for advice to help their own gorillas breed successfully, and the *Today Show*, *Time* and *Life* magazines, and the *New York Times* all carried stories about Colo. When old enough to be put on view, over a million people showed up in 1957 to get a peek at the little one in the nursery, often bedecked in frilly outfits. Scientists came to study her development, and other zoos offered large sums to borrow Colo, even for just a little while. All the offers were turned down. Her human baby–like treatment came to an end when she snuck up on a keeper feeding a baby bear. A cage would thus serve as her new home.

Later, Colo became pregnant from Bongo, a wild-born gorilla, who at nineteen months of age was purchased specifically to be her mate. They had bonded quickly and with maturity began mating. The happy event occurred on February 2, 1968, to be followed by another on July 18, 1969, and yet another on December 28, 1971, her last. To add to the list of firsts, Colo became the first captive gorilla to achieve the status of grandmother.

Colo is still going strong and is the world's oldest known living gorilla. In all, there are five generations stemming from her represented at the Columbus Zoo. And in Colo's golden days, she has become a bit of a prognosticator, having picked the University of Connecticut Huskies as the Final Four winner in the 2011 NCAA basketball tournament. Her success led *Good Morning America* to show up on March 30, 2012, to showcase Colo using numbered turnips to predict the winner of the huge Mega Millions lottery. This time she didn't fare as well by picking 9, 12, 21, 31, 41, and Megaball 9 instead of the winning combination of 46, 23, 38, 4, 2, and Megaball 23.

On September 23, 1959, at the Basel Zoo, Goma became Colo's European counterpart. This was a planned event for mother Achilla and father Stephi, the male obtained from the Columbus Zoo in 1954. At the time, they were the only gorilla pair on the continent. Although Achilla held her newborn tenderly, she, like Millie, had no clue about how to nurse, so zoo director Ernst Lang took Goma home, where she remained for two years to be raised pretty much like a child. Later, he wrote a short book called *Goma, the Baby Gorilla*, describing her care and development. Here's what he said about Goma's social needs:

> When we go to lunch, Goma must be there, for as we had to keep reminding ourselves, gorillas in the wild live in families and tribes, always with plenty of company, and if a young gorilla were left alone too much it would not flourish. So we offered our Goma as many opportunities for contact as possible—and she is happiest when our whole family is together.[25]

Then one evening, he noticed that

> she was playing with the lowest branches of a shrub, pulling them down, breaking them off, putting her foot on them, tearing the fibres out one by one and happily beating the leaves. Suddenly she began to arrange the branches near her, in the way chimpanzees do when they want to build a nest on the ground. As darkness began falling she came over to me on the garden seat, but turned back to the shrub to arrange the branches. When it became quite dark she came back and lay in my lap.[26]

The media jumped on Goma's birth, and within a few weeks she had become a global sensation. Although the public clamored to see her, Lang waited until the zoo held a big coming-out party on her first birthday. Since then, Goma's peaceful demeanor has drawn visitors to the Basel Zoo, and on September 23, 2009, a grand fiftieth birthday celebration was held in her honor.

One would not have predicted stardom for Timmy, or Tiny Tim, as he was first called upon arrival at the Memphis Zoo in 1960, an orphaned one-year-old from Cameroon. There he sat simply to be looked at by people in the then common concrete and steel cage housing gorillas. Matters changed when in 1966 the Cleveland Zoo acquired him. In better surroundings, Timmy responded by being more active and outgoing, and during a twenty-five-year stay he developed a substantial following among local residents. He did not, however, care much for the several females brought in for breeding purposes until Kate arrived. They bonded, but she proved to be infertile, ending the zoo's hopes for a family.

Given his born-wild pedigree, the Cleveland Zoo agreed to loan Timmy to the Bronx Zoo in order to diversify its gene pool: all of its females had been born in captivity. Animal activists and zoo supporters in Cleveland took to the streets and wrote letters protesting that the move could be detrimental to Timmy's health by taking him away from his "love," Kate. They filed suit but lost, and off he went to the Bronx Zoo in November 1991. Timmy didn't have much success at first, but he eventually adapted to his new environment and fulfilled a silverback's dream by siring thirteen offspring. When his breeding days came to an end, he found a new home in 2004 at the award-winning "Gorilla Forest" of the Louisville Zoo, where people flocked to admire the handsome silverback. When Timmy turned fifty, he was serenaded with "Happy Birthday," led by none other than President Obama. The end came on August 1, 2011, when his chronic heart conditions and arthritis prompted zoo officials to euthanize him in order to spare him further suffering. At fifty-two years of age, Timmy was the longest-lived male gorilla in North America when he passed from view, an event that brought forth an outpouring of tears and eulogies from his many fans in Louisville, in Cleveland, and across the country. As for Kate, she lost a toe in a fight with Timmy's replacement, Oscar, and wound up at the Fort Worth Zoological Park.

In 1964, Rio Muni farmer Benito Mañé shot and killed a female gorilla raiding the family's banana plantation. To his utter amazement, he found a young one covered with white hair clinging to her back. No white gorilla had ever been seen before, so Mañé, sensing that he possessed something of value, cared for the *Nfumi Ngi* ("white gorilla") for several days before selling him to Jordi Sabatar Pi. After a short period of conditioning to life in captivity at his special facility in Rio Muni, Sabatar Pi shipped the youngster to the Barcelona Zoo. There he became known as Little Snowflake (*Floquet de Neu* in Catalan), later shortened to Snowflake. He spent his first eleven months at

Plate 6.4. Snowflake playing the role of a silverback at the Barcelona Zoo. *Source:* Barcelona Zoo Photo Archive.

the apartment of the zoo's veterinarian Dr. Romá Luera Cabó, whose wife did much of the care and feeding. Once he was put on display, people flocked to see the white-haired little creature that enjoyed playing for an audience. With blue eyes, Snowflake wasn't a true albino, but he did have daytime vision difficulties that probably would have led to an early death in the wild.

The National Geographic Society got wind of Snowflake, and the two articles that appeared in its magazine in 1967 and 1970 quickly turned him into an international celebrity.[27] Postcards carried his image, and tourist guidebooks recommended visits to the Barcelona Zoo specifically to see him. And it didn't stop there. *Nature* produced an episode called "Snowflake: The White Gorilla," and later two books were written to commemorate his life—*Memoires d'en Floquet de Neu* (2003) by Toni Salai and *Floquet per Sempre* (2003) by Sabatar Pi. The announcement that Snowflake was afflicted with terminal skin cancer resulted in thousands coming to say their final farewells. His last day came on November 24, 2003, after having entertained multitudes and siring twenty-two offspring, none, though, with white hair. To this day, Snowflake remains the only real white gorilla to have been documented, the others being fictional.

On August 16, 1996, at the Brookfield Zoo in Chicago, the spotlight shined on Binti Jua. Among the assembled crowd, a rambunctious three-year-old boy started to climb over a barrier to get a better look at the apes located in an enclosure below. He suddenly lost his balance and fell some eighteen feet to the hard concrete floor. Binti Jua was sitting nearby and wandered over to have a look. She carefully picked up the limp body and slowly walked nearly sixty feet to deposit the boy at the entrance door to the exhibit, where attendants could remove him. Binti Jua then returned to where she had been sitting.

In today's language, the story quickly went viral. Thousands of congratulatory letters poured into the zoo, with Binti Jua a heroine in many people's eyes. She even received a medal from the American Legion. Naturally, the incident raised the issue of what caused Binti Jua to behave as she had. Some people thought it an act of pure altruism. More likely, motherly instincts came into play. She had been hand raised at her initial home in the San Francisco Zoo and was thus familiar with people, and at Brookfield she had been taught how to be a gorilla mother and, indeed, was one at the very moment, the infant clinging tightly to her back during the whole episode. Afterward, Binti Jua went back to being a normal zoo gorilla. And it should be noted that the boy recovered fully after a few days in the hospital.

The best-known gorilla outside circuses and zoos has to be Koko. Indeed, she likely qualifies as the most famous real gorilla ever, surpassing even Gargantua. Her rise to fame began when Francine (Penny) Patterson visited the San Francisco Zoo in October 1971 and saw a three-month-old female Western gorilla named Hanabi-Ko, shortened to Koko, trying (but without much luck) to hold tight to her mother. Patterson was a PhD student in developmental psychology at Stanford University and had experience working with primates. She also knew about the promising research being done by Beatrice and Alan Gardner to teach American Sign Language (ASL) to the chimpanzee Washoe. In the 1950s, simians had come back to center stage in the animal communications debate. After attending a talk by the Gardners, Patterson decided to undertake a similar experiment with Koko. At first denied the opportunity, she got her chance after Koko had been transferred to the Children's Zoo following a serious illness. Patterson visited every day in order to establish a bond between them and slowly began to teach Koko simple signs.[28] After a month, she reportedly knew the one for "food," and others, such as "dog" and "drink," quickly followed.

Patterson wanted more privacy for the lessons and found a ten-by-ten-foot used mobile home that would do the trick. As the months progressed, she decided that a less distracting environment with more frequent contact was also needed, and Patterson approached Stanford University authorities about relocating the mobile home to the campus. In September 1974, her wish came true, and two years later she claimed that Koko knew 200 words of "Gorilla Sign Language."

Patterson then found a partner for Koko, Michael, whose mother had been killed in Cameroon by hunters looking for bush meat. Soon after arriving, Michael reputedly began signing as well, Patterson later saying the two gorillas learned to communicate with each other via signs.

Koko's future remained in doubt. Would she remain with Patterson or be returned to the San Francisco Zoo? In the summer of 1977, negotiations allowed Patterson to buy Koko for her nonprofit Gorilla Foundation. Longtime partner Dr. Ronald Cohen served as cofounder, primary on-site photographer, and video expert. Better financing allowed the Gorilla Foundation to find Koko and Michael a new home in a wooded area of the Santa Cruz Mountains near Woodside in October 1979. By this time, Koko supossedly could sign over 500 words. Eventually, the reported total would exceed 1,000, and, according to Patterson, she knew even more spoken words.[29]

A ten-year old named Ndume, obtained from the Cincinnati Zoo, joined the family in 1991. Michael and Koko hadn't mated and the Gorilla Foundation hoped for better things with Ndume, as he had produced offspring in Cincinnati. This, too, proved disappointing, especially coming on top of Michael's unexpected death in April 2000.

During all this time (and even before), the issue of the language learning capabilities by great apes had come under fire. The heat intensified when allegations surfaced that Washoe didn't know ASL after all. The lone deaf member on the team—and thus the only one intimately familiar with the language—said that a number of the signs others recorded seeing were not recognizable to him.[30] As it turned out, many of Washoe's signs did not, in fact, match with those of ASL. Instead, these consisted of specially created ones formed by molding her hands in a particular way to represent specific objects or involved other signs that she created herself.

To find out if Washoe knew how to use signs to create a language, Herbert Terrace at Columbia University decided to work with another chimpanzee, whimsically named Nim Chimpsky after the famous linguist Noam Chomsky. Terrace had his doubts about the claims being made, believing that Washoe might be merely responding to signs initiated by teachers and that tricks and inducements had been used to get correct responses from her and from other chimpanzees that the Washburns had trained.[31] Terrace thus initiated Project Nim, with the intent of socializing

> a chimpanzee so that he would be just as concerned about his status in the eyes of his caretakers as he would be about the food and drink they had the power to dispense. By making our feelings and reactions a source of concern to Nim, I felt that we could motivate him to use sign language, not just to demand things, but also to describe his feelings and tell us about his views of people and objects. I wanted to see what combinations of signs Nim would produce without

> special training, that is, with no more encouragement than the praise that a child receives from its parent. I especially wanted to find out whether these combinations would be similar to human sentences in the sense that they were generated by some grammatical rule.[32]

Nim learned 125 signs over the course of the project's forty-four months and produced some combinations. He also learned many other things, like washing dishes and doing laundry. In the end, though, Terrace concluded, "For the moment, our detailed investigation suggests that an ape's language learning is severely restricted. Apes can learn many isolated symbols (as can dogs, horses, and other nonhuman species), but they show no unequivocal evidence of mastering the conversational, semantic, or syntactic organization of language."[33] Still, he held out hope future efforts might well produce a chimpanzee that "understood the power of signs as a means of communicating about its world."[34]

In 2011, HBO and the BBC released a documentary titled *Project Nim*. The first half contains interesting footage of Nim's training, interspersed with interviews of the various people involved, including Terrace, who comes across as a somewhat cool and distant person. The last half deals with Nim's life after being returned to the Institute for Primate Research in Oklahoma, from which he was obtained when two weeks old. It's disturbing to watch the conditions that Nim endured there and afterward at animal rights advocate Cleveland Amory's refuge Black Beauty Ranch in Texas, where he was the lone chimpanzee for a considerable period of time until some others from research facilities arrived. In May 2000, Nim died of a heart attack at the age of twenty-six.

Psychologists David and Ann Premack came at the subject of ape language learning from a slightly different perspective, thinking that it might tell them more about human language development. To this end, they invented a special language in which each word was represented by a piece of differently shaped colored plastic. The study consisted of nine subjects, the most famous being their first, Sarah, who began her lessons in 1967. She and all the other chimpanzees in the study were wild born and raised in comfortable settings (although not in homes) and given several tests per day by specially trained instructors. Over the course of these sessions, Sarah and her cohorts learned many things and demonstrated capacities for imagery and abstract representation. The minds of these chimpanzees had clearly been enriched beyond what they had been at the beginning, and, according to the Premacks, they can with effort be taught a language. But as David Premack notes, "Teaching language to this sort of mind does not confer human language," and furthermore "adding a human larynx to the ape would not make of it a human."[35]

In the 1990s, Steven Pinker entered the fray. A rising star in experimental psychology and cognition, he argued in his popular book *The Language*

Instinct that no ape had learned ASL and that any signs they did make after long hours of training "were not coordinated into well-defined motion contours of ASL and were not inflected for aspect, agreement, and so on—a striking omission, since inflection is the primary means in ASL of conveying who did what to whom and many other kinds of information."[36]

Noam Chomsky himself had long disputed the findings put forth by the ape projects. In a 1980 *Time* magazine article, he dismissively remarked, "It's about as likely that an ape will prove to have a language ability as that there is an island somewhere with a species of flightless birds waiting for human beings to teach them to fly."[37] Later, he said that no matter what the tests showed, the subjects were merely picking up on human clues, in a sense giving their instructors what they want, and that the claims made on behalf of language learning by great apes are driven by "sentimentality," not science.[38] In a review of the Washoe project and related efforts, Yale University linguist Stephen Anderson concluded,

> The Washoe Project suggested strongly that it is possible to teach chimpanzees a substantial vocabulary of arbitrary signs, in the form of manual gestures with an associated meaning that is at best only partially related to the form of the gesture itself. Little or no evidence exists for any linguistic structure beyond this, and certainly none for full (or even substantial) command of human language.[39]

The most vocal advocate for the language learning abilities of great apes is Sue Savage-Rumbaugh, who, along with her husband, Duane, began working with chimpanzees and bonobos at Georgia State University's Language Research Center in 1971. They've used lexigrams and computers rather than sign language and assert that language learning complete with grammar and syntax is possible with both species, although not all subjects have shown this ability. Kanzi, a bonobo, has been their most accomplished student. He is said to have learned to communicate by watching his mother, who failed to catch on, and accomplished some remarkable things in stringing words together and recognizing spoken words. But the question of his responses possessing grammar and syntax is still an open one. Then there's the fact of the subjects being raised with intense human interaction from an early age. This is vividly portrayed in *Apes, Language, and the Human Mind.*[40] They are, quite literally, no longer chimpanzees and bonobos, at least with regard to natural behavior.

The controversy over language learning by apes has yet to be settled largely because there is no generally accepted agreement on what language is, and it's doubtful that agreement will be reached any time soon. In *Chasing Doctor Dolittle: Learning the Language of Animals,* biologist Con Slobodchikoff goes back to Garner and even beyond him by essentially equating all types of communications, such as bee dancing and color changes in male squid, as languages of a sort. He sees it as embedded in evolution and not a "gift granted only to humans."[41] What he doesn't do, however, is address whether human language is fundamentally different from all other forms of communication, which is the concern of many linguists and behavioral psychologists. However, the highly technical nature of the arguments makes it difficult for nonspecialists to come to any kind of informed judgment on the issue.[42] Slobodchikoff admits that learning the language of another species is extremely time consuming and difficult and communicating with them in it even more so. Take the case of humans and our companion animals. A lot of communication goes on via sounds and gestures to the point where often we seem to understand one another. They can clearly learn what we mean by "no," but what is passed on to them by mimicking meows and barks? The same can be said about Dian Fossey's learning to utter some sounds of her gorilla subjects for purposes of habituation. In our current state of understanding, I think all one can confidently say is that her efforts seem to have comforted them.[43]

As for the Koko project, there are features peculiar to it that invite skepticism. A key one relates to data. Patterson has not shared as much as might be desired. Furthermore, the project is the only effort so far to work with gorillas in an intensive manner, and thus no comparisons with claimed results are possible. In addition, some statements made can cause eyebrows to rise. One involves reports about Koko's understanding complete sentences. For instance, when she wouldn't sign "rock" accurately after being told to do so several times, Patterson said, "I won't give you your night dish unless you say 'rock.'" Koko then reportedly signed "rock."[44] If true, then she does communicate with an understanding of grammar and syntax. However, we have only Patterson's word to go by. Perhaps the most questionable assertion made on behalf of the Koko project is that one day out of the blue, Michael signed a detailed account of his mother's being killed by hunters.[45] He was an infant still clinging to her at the time this happened!

One thing isn't in doubt, however, namely, that Patterson via Koko has brought a lovable gorilla into millions of human homes. They've appeared together on such television programs as *Nova*, *Mr. Rogers' Neighborhood*, *Dateline*, and *48 Hours* and featured in stories included in *National Geographic*, the *New York Times Magazine*, and the *Wall Street Journal*. Perhaps even more important in changing perceptions about gorillas are the children's books Patterson has written. These are full of endearing photos and stories showing Koko as being naughty at times, tickling, playing tricks, loving kittens, playing with dolls, laughing, and walking hand in hand. Some may see Patterson as taking anthropomorphizing a bit too far, but, as George Schaller noted, this may be impossible not to do with gorillas and, one might add, also necessary for their survival.

The Gorilla Foundation remains in its Santa Cruz home, awaiting a new one at the Maui Ape Preserve located on the west side of the island in a seventy-acre tropical setting. Designed for education and "interspecies communication," its groundbreaking ceremony took place on October 12, 2000; however, a series of complications kept it from opening. In 2007, a decision was reached to reconsider the design and purposes of the preserve. Should it ever come to fruition, will Koko live long enough to enjoy the surroundings?

The time has come to shift the scene to that other source of important information about gorillas—movies—to see what they have communicated.

ON THE SILVER SCREEN

Monster movies became all the rage in the early days of cinema. *Balaoo* (1913) featured a human changing into a rampaging baboon, and Lon Chaney played a sinister doctor who transformed a young man into a monster in *A Blind Bargain* (1922). *Ingagi* (1930) is a classic exploitation film.[46] It purportedly told the story of a British expedition to Africa that came across a tribe worshipping gorillas and offering its virgin women to them in sacrifice. The gorillas naturally obliged by carrying off the prizes to jungle homes to satisfy their sexual desires and apparently those of the women as well. Crowds flocked to see the film, but it was soon banned from regular theaters when the whole story proved to be a hoax. Much of it had been made in Los Angeles with African Americans playing the roles of Africans. *Ingagi* did, however, continue to be shown at art houses and other venues for years afterward.[47]

The movie *King Kong* arguably gave rise to the most well known gorilla of all time. Kong first appeared on the silver screen in 1933, and while not an instant star, his legend grew sufficiently to produce rereleases in 1938, 1942, 1952, and 1956, with the latter two especially successful because of the growth of drive-in theaters catering to family and teenage audiences. In 1976 and 2005, remake introduced the gigantic gorilla to new generations, and together the movies helped create a small industry of scholarship, ranging from production issues to inquiries into the movies' social and cultural meanings.[48] My concern here is with the less well covered issue of how the movies represented gorillas to viewers and how these representations related to the information available to their creators. Because it set the mold, attention will be focused on the 1933 version, with the later productions looked at primarily in terms of their differences from the original.

King Kong came into being via the creative minds of the travel/documentary team of Merian Cooper and Ernest Schoedsack, aided by screenwriter Ruth Rose and animator Willis O'Brien. They wanted to make an unparalleled

Plate 6.5. Poster for the exploitation movie *Ingagi*. *Source:* Photofest.

adventure story centered on an awesome beast. The initial working title of the film was, in fact, *The Beast*, and during the early stages of production Cooper used *The Eighth Wonder*. Eventually, the subtitle for the film became *The Eighth Wonder of the World*.

Kong's home is situated in the depths of a tropical forest on Skull Island, located somewhere near Sumatra in Southeast Asia. The notion of such forests harboring strange and dangerous beings hadn't gone away, and rumors kept cropping up about the existence of uncharted islands where creatures dating to prehistoric times could be found. The discovery of the Komodo dragon in 1910 served to validate the possibility. Dinosaurs and other prehistoric animals had been featured in previous movies, most notably *The Lost World* (1925), but rather than repeat the scenario, Cooper came up with the idea of making a giant prehistoric gorilla the centerpiece of the story, one to be, in his words, "fifty times as strong as a man—a creature of nightmare horror and drama."[49] The fact that gorillas lived only in Africa didn't matter. This wasn't meant to be a documentary. More important, they actually existed, and, furthermore, what most people knew of them emphasized size, strength, and fierceness. All that needed to be done to achieve the "nightmare horror and drama" goal was to make these traits larger than life.

The RKO production clearly accomplished its objective, as illustrated by the screams from audiences matching those by Fay Wray in the role of Ann Darrow. The "nightmare horror" first appears when a monstrous and hideous-looking Kong ravages a native village and then steals away with Ann, his beautiful blond bride. In *Ingagi* fashion, the villagers had been offering their own young women to propitiate him. Kong takes Ann to his mountain lair and at one point tears her clothes, a clear reference to the old story of gorillas sexually lusting after women. Her brief exposure was quickly deleted from the movie after initial showings. Kong also demonstrates his might by killing an array of prehistoric creatures threatening the prize he has claimed. The combat completely exhausted him, and thus, when the film's handsome hero Jack Driscoll (Bruce Cabot) and not scheming promoter Carl Denham (Robert Armstrong) arrives on the scene, he's able to spirit Ann away to safety. Later in the movie, Kong again demonstrates his ferocity when he escapes from captivity in New York City and goes on a rampage, first turning over an elevated train. After this, a squirming man is seen dying between his teeth, and in another scene he grabs a man attempting to flee from a car accident and flings him into a crowd. Horrifically, Kong then snatches a sleeping woman from bed and, upon discovering that she is not Ann, lets her fall to her death. Such scenes, along with the posters meant to lure people into theaters, certainly played upon and reinforced prevailing views of gorillas as beasts.

Plate 6.6. One of several posters for the 1933 version of *King Kong*. *Source:* Photofest.

Kong, however, displayed another side. At his lair, it's not clear if the battles fought are to keep his prize for himself or to protect her. But as events proceed, his gentle handling of Ann reveals that Kong's motives involved both possessing and protecting her. As such, he's something more than a beast, and to accentuate his human side the producers have him box and wrestle with foes in human style. This theme is played on in New York City when Kong finally does find Ann and carries her to the top of the Empire State Building. Again, we're not sure what he has in mind, although it soon becomes clear that he loves her. Nonetheless, as all who have seen the movie know, his love proved to be in vain, for after absorbing uncounted bullets from attacking airplanes, Kong falls to his death.

A question arises about where the filmmakers got their information about gorillas. Cooper claimed to have read Du Chaillu as a youngster, and this must have had some influence on his thinking. But was he or someone else on the team familiar with *Captured by a Gorilla*? It's not listed in the film credits, but Kong's stealing away a beautiful woman and her later rescue by a team member is remindful of the story. Some or all of the film team must have known about Carl Akeley's exploits, and most certainly Cooper and Schoedsack had seen *Gorilla Hunt* by Ben Burbridge as well as *Man Hunt*, a 1926 production from FBO, RKO's precursor, recounting Burbridge's exploits. During the making of *King Kong*, the Johnsons' *Congorilla* hit theaters. Without credits or other information, one can only speculate. Still, whether intended or not, the movie matches a moment in time when new information concerning gorillas had begun to change the way people would think and talk about them. This is highlighted by Carl Denham's classic line at the end: "Oh no, it wasn't the airplanes. It was Beauty killed the beast." This brings the story full circle, for at the beginning we hear, "And the beast looked upon the beauty and lo! his hands were stayed from killing. And from that day forward, he was as one dead." But in Cooper's mind, *King Kong* was more than a beauty-and-the-beast fable. As he later noted, "If I can get that gorilla logically on top of the mightiest building in the world and then have him shot down by the most modern weapons, the airplane, then no matter how giant he was in size and how fierce, that gorilla was doomed by civilization"[50]—proved prophetic, as events since have shown.

Nine months after *King Kong* debuted, RKO released *Son of Kong*. The story line has Carl Denham setting out on another adventure in search of big profits, and eventually he and his party wind up back on Skull Island, where they meet up with a much smaller albino gorilla about twice the size of a man. He's friendly from the outset and saves Denham and a girl from a giant cave bear and later dispatches a nothosaur bent on killing them. Called both "Little Kong" and "Baby," he and all the other inhabitants of the island die when it sinks beneath the sea following an earthquake, while the human stars get away, with thanks owed to Baby. The movie was only a modest box office success and never developed a substantial following.

The 1976 Dino de Laurentiis remake of *King Kong* featured a different cast of characters (Charles Grodin as oil company shark Fred Wilson, Jessica Lange as aspiring actress Dwan rescued at sea, and Jeff Bridges as paleontologist/photographer Jack Prescott), but otherwise the story pretty much mimics the original. Kong is the same roaring god beast of the inhabitants on Skull Island; Dwan is captured by them and set out to be Kong's latest bride;

Jack rescues her; Fred designs a way to capture Kong, who's taken to New York City to star with Dwan in a spectacle; Kong escapes and begins ravaging the city; and he finds Dwan and climbs the Twin Towers with her, this time to fall to his death after being shot multiple times by helicopter gunships. There is one difference, however. Dwan's love for Kong parallels his for her. She saves him from death on the ship, carrying all of them back to New York City, and stays by his side trying to rescue him as he's being shot, and on the pavement the two have a last longing glance at each other as his eyes slowly close. Kong thus becomes even more humanlike than the one created by Cooper and Schoedsack, and consequently his passing seems more like a crime, a point that Jack makes several times during the movie.

This version of *King Kong* opened and closed rather quickly at theaters. The story had been told before; it was campy more than scary and had Dwan uttering a number of corny lines. It did eventually come out on DVD, but there doesn't seem to have been a rush to buy or rent copies.

In 1986, the much-delayed sequel *King Kong Lives* appeared. As the title implies, Kong didn't die after falling from the World Trade Center. The implausible rescue story and his eventual death led to a film so bad that it disappeared even more quickly from theaters than its predecessor, never to be seen again except by those interested in filmography or gorilla portrayals.

This brings us to Peter Jackson's epic three-hour-plus 2005 remake. Like the original, it begins in Depression-era New York City and ends there after the journey to and from Skull Island. Kong doesn't appear until about two-thirds into the movie. He's huge, of course, but has more of a silverback's physical features than the 1933 version. And while roaring and showing his massive teeth when taking Ann Darrow (Naomi Watts) to his lair, seemingly bent on tearing her from limb to limb like previous sacrificial brides, in many ways he's not nearly as scary as the zombielike inhabitants of the ruined ancient city on the island. Furthermore, Jackson introduces all kinds of prehistoric monsters from dinosaurs to spiders to bats that make Kong look relatively benign. Upon reaching the lair, his nature quickly changes. He appears old and weary from many battles, a kind of "melancholy hero," according to Cynthia Erb.[51] Kong is, after all, the last of his line, a fact made more powerfully in this rendition of the film than the other two.

During their time together at the lair, the love story begins, one that intensifies as the movie progresses, although without sexual connotations, until Kong meets his doom, again by falling from the top the Empire State Building after an attack by airplanes. Two scenes leading up to this point are particularly poignant about Kong. After finding Ann during his rampage through Manhattan, they playfully scoot around the ice in Central Park until interrupted by soldiers come to kill him. And at the top of the Empire State

Building, they watch a sunrise together, as they had once watched a sunset on Skull Island, hand in hand. Kong even utters the word "beautiful." He then saves her one last time. To borrow a phrase from Rene Dubois, at the end Kong becomes "so human an animal." Jackson, clearly, had done his homework in putting the various aspects of the gorilla star together.

After a lull during World War II, movies featuring gorillas began appearing again at its end. *White Pongo*, released in 1945, is actually rather interesting. It revived the old word for gorilla, was said to take place in the Congo, and anticipated Snowflake. In addition, stories claimed White Pongo to have near human intelligence and to be the likely "missing link." Thus, an expedition is mounted to find the creature, and when this happens, viewers see an ape approximating the size of an actual gorilla but with a terrifying face nowhere near matching the real thing. Borrowing from *King Kong*, White Pongo steals off with the only woman in the expedition, takes her to his lair, wards off a lion, and battles a monstrous black gorilla that he subdues with a tree limb. The others in the expedition, of course, rescue the woman and capture White Pongo. The film ends as he's about to be shipped to London for exhibit.

Reality took a hard hit with the 1951 release of *Bride of the Gorilla*. It's set on a plantation in the Amazon, an early giveaway of what's to come, and tells the story of a man who at night turns into a gorilla after being administered a potion by a local priestess. He had killed the boss of the plantation to get his wife and then marries her. Only a few glimpses are seen of a hairy biped running through the jungle. In classic style when killed, the gorilla slowly changes back into a man. Interestingly, Lon Chaney Jr. is in the movie, not as the man/gorilla but rather as a police officer familiar with the jungle and responsible for tracking down the creature.

Arguably the worst gorilla movie of all time and maybe close to the worst movie period is the 1968 production *Kong Island*. It has nothing to do with *King Kong* per se, and there's no island. Rather, the setting is a supposed jungle in Kenya where a mad scientist has created hideous-looking robotic slave gorillas that walk fully upright. Naturally, one of them steals away with a young woman, who's part of a group of American adventurers seeking to find a "sacred monkey." There is, of course, also a hero, and to end the story quickly he destroys the device controlling the brains of the gorillas. The woman is saved, and the gorillas wreak revenge on their master. The "sacred monkey" turns out to be a Jane-like person living in nature, and she's seen at the end of the film happily running into the forest with a chimpanzee friend.

Gorillas would appear in a number of other movies over the years, most notably in *The Planet of the Apes* and its various successors. They also make appearances in the numerous Tarzan movies, although gorillas were never referred to as such. They became just apes. And there's King Kong's battle with Godzilla. The only other gorilla, however, to achieve some degree of stardom is *Mighty Joe Young*, who is featured in two films, the first in 1949 under guidance of the same team behind *King Kong* and *Son of Kong*. Joe, though, is portrayed as a more believable gorilla. Viewers first see him in the guise of a real baby taken in by a young daughter of a plantation owner in some unidentified part of Africa. He's put in a crib, and she treats him as if he were a little brother. When next seen, Joe is fully grown, and, while huge, he's not monstrous and possesses none of the hideous facial characteristics of Kong or other previous gorillas. In a bit of realism, Joe does some knuckle walking. Even in the two action episodes, he's not the bad guy. In the first one, Joe fends off a bunch of cowboys seeking to capture him in what can only be described as one of the silliest movie scenes ever. Not a single rider or horse is killed or, for that matter, even injured. In good little-brother fashion, his friend, now a teenager played by Terry Moore, is able to calm Joe down by singing "Beautiful Dreamer," a song she had used to comfort him when he was a baby.

Joe later winds up in Hollywood to star with her in a nightclub act set in a pseudo-jungle environment where he's billed as Mr. Joseph Young. To prove his strength, Joe bests ten of the world's supposedly strongest men in a tug-of-war. Soon, though, he becomes depressed, as gorillas were wont to do in captivity, and during the seventeenth performance of the act confusion and anger set in when the gawking audience starts throwing fake coins onstage. Back in his cage, three drunks ply Joe with alcohol, after which he starts to tear the nightclub apart. Again, no one gets killed, the only casualties being some lions that escaped during his rampage. Joe has our sympathy throughout the ordeal, as people are responsible for what is happening, not him. While being hustled away to escape a shoot-to-kill order, Joe becomes a hero by rescuing a young girl from a burning school building, risking his own life in the process. Unlike *King Kong*, the film ends happily with Joe back in Africa, accompanied by his friend and her newfound love together in a scene of domestic bliss. Clearly, the gorilla of Akeley has won out over that of Du Chaillu. Although the movie won an Oscar for special effects, it failed at the box office, and a planned "Joe Meets Tarzan" follow-up was scrapped.

Still, Joe came back into being in a 1998 remake. Real gorillas are shown at the beginning of the movie in an again undefined part of Africa, reminiscent, though, of Rwanda. In addition, there's a tough Dian Fossey–like woman named Ruth Young who saved a baby gorilla from a white hunter who killed its mother. The baby had bitten off a thumb and finger of the hunter during an attempt at capture. Afterward, the hunter kills Young. As in the previous movie, viewers next meet Joe twelve years later being taken care of by Young's daughter, Jill (Charlize Theron). He's grown to a height of fifteen feet, ten less than Jackson's gorilla, and weighs a ton. Still, he's an outcast except for Jill. Onto the scene comes Gregg (Bill Paxton), a director of a wildlife refuge. Poachers have arrived, and Gregg convinces Jill that for safety's sake Joe should be relocated to the refuge he's created in Hollywood. The evil white hunter has learned about Joe's presence and wants revenge for the damage inflicted on him earlier. Later, at a party held on the refuge, one of the hunter's cronies blows a noisemaker used by poachers. Joe goes on a rampage and is pursued by the police in cars and helicopters. Jill and Gregg eventually find Joe at an amusement park where he's raising havoc but threatening no one. The hunter, too, is there and sets out to kill Jill. Joe sees this and tosses the man onto electrical wires, where he's electrocuted to the delight of audiences. Meanwhile, the police are poised to shoot Joe. They're convinced to hold their fire when Joe starts climbing a Ferris wheel that has caught fire in order to rescue a young boy stranded at the top. He succeeds in grabbing him but then falls. Both survive, and Joe is returned to Africa, where he lives free in a sanctuary established with donated funds for his courageous rescue, with Jill and Gregg present to see him run happily off into the forest.

Despite lacking a gorilla star and with humans costumed as gorillas, except for a few spliced-in scenes, the 1999 movie *Instinct*, based loosely on the novel *Ishmael* by Daniel Quinn, belongs in the pantheon of Hollywood's portrayal of gorillas. The story centers on an anthropologist (Anthony Hopkins) confined to a maximum-security prison for the mentally ill. He's violent and doesn't speak until a psychologist (Cuba Gooding Jr.) manages to break through his wall of silence. Their conversations reveal that the cause of the anthropologist's behavior resides in a relationship established with Mountain gorillas he had been studying. Seduced by their peaceful, natural life, he intensifies the contacts and slowly becomes accepted into a family, abandoning research in the process. One can see Dian Fossey lurking in the story. The peace is shattered when, for reasons left unspecified, park rangers begin shooting members of the family. Seeking to protect them, the anthropologist goes after the hunters and in the process kills two of them and wounds several others. He is, in effect, playing the role of a silverback. The remaining rangers manage to subdue him, and, seeing this, the real silverback comes to the rescue, only to be shot dead, thus bringing on the psychosis. Cured by help from the psychologist, he's about to go before the parole board when an incident by a violent guard against his new family, the other inmates, causes him to come to their defense. Denied the hearing, the anthropologist becomes silent again. But it's a ruse, and with the help of the inmates he escapes and is last seen returning to the forest, presumably in search of a peaceful life with the gorillas. A number of other themes run through *Instinct*, including control, people as "takers," human dominion over the earth, and freedom. There's also a scene criticizing zoos for turning gorillas into mere shadows of their real selves. Not a popular film at the box office, it did win a Genesis Award for animal rights by showing humans as the perpetrators of violence and depicting gorillas as they really are.

As the twentieth century gave way to the twenty-first, popular culture had pretty much caught up with science in its image of gorillas. Yes, the past does hang on to some extent. An example is the "Curse of the Silverback" water slide at the Enchanted Forest Water Safari Amusement Park in Old Forge, New York. Actually, there's nothing scary about the silverback standing over riders as they swish by. It's the speed through the swirling water that provides the thrill. Then there are two recent cartoons in the *New Yorker*. One shows a huge silly-looking ape licking his chops from behind a building as a woman walks by on the street below, whereas the other has a King Kong–looking gorilla climbing a building and saying to a blond woman held in his hand, "Mate with you? I was just going to use you as a Q-tip."[52] If anything, strength now comes to mind, as witnessed by Gorilla Glue, Gorilla Tape, and Gorilla Glass. And, for many people today, the image of them being "gentle giants" has come to be a more proper representation. But do they have a future on the planet? This will be our final consideration.

NOTES

1. Bits and pieces of John Daniel's story can be found in Alyse Cunningham, "A Gorilla's Life in Civilization," *Bulletin of the New York Zoological Society* 24 (1921): 118–24; *New York Times*, April 13, 1921; and Anonymous, "John Daniel, Civilized Gorilla," *Literary Digest*, December 10, 1922, 44–49.
2. Thomas Alexander Barns, *Across the Great Craterland to the Congo* (New York: Alfred A. Knopf, 1924), 133–34.
3. Richard D. Sparks, "Congo: A Personality," *Field and Stream*, January 1926, 8–20, 72–73.
4. Ben Burbridge, *Gorilla: Tracking and Capturing the Ape-Man of Africa* (New York: The Century Co., 1928), 286–87.
5. Robert M. Yerkes, "The Mind of a Gorilla," *Genetic Psychology Monographs* 1927: 33.
6. Ibid., 180–81.
7. Roger Conant, "Bamboo," *Fauna* 1 (1939): 7–9.
8. Roger Conant, "Meet the Champions," *Fauna* 3 (1941): 48.
9. For information on Madame Abreu, see Robert M. Yerkes, *Almost Human* (New York: The Century Co., 1925), and Ramona Morris and Desmond Morris, *Men and Apes* (New York: McGraw-Hill Book Co., 1966).

10. Mark Rosenthal, Carole Tauber, and Edward Uhlir, *The Ark in the Park: The Story of Lincoln Park Zoo* (Urbana: University of Illinois Press, 2003).

11. D. B. Willoughby, *All about Gorillas* (Cranbury, NJ: A. S. Barnes and Co., 1979), 180.

12. The little that is known about Susie can be found in Caroline Dressman, *A Brief History of "Susie" as Told by Her to Her Trainer Wm. Dressman*, a tiny and rare publication held by the Cincinnati and Hamilton County Public Library and the Cincinnati Historical Library and Archives, Cincinnati Museum Center.

13. Gertrude Davies Lintz, *Animals Are My Hobby* (New York: Robert McBride & Company, 1942), 106.

14. Ibid., 118.

15. Henry Ringling North and Alden Hatch, *The Circus Kings: Our Ringling Family History* (Garden City, NY: Doubleday & Company, 1960), 266–67.

16. Lintz, *Animals Are My Hobby,* 195–96.

17. J. Y. Henderson, *Circus Doctor* (Boston: Little, Brown and Company, 1973), 209.

18. A. M. Hoyt, *Toto and I: A Gorilla in the Family* (Philadelphia: Lippincott, 1941), 150.

19. North and Hatch, *The Circus Kings*, 272.

20. Illustrations of Mr. And Mrs. Gargantua, along with an array of photographs of the two gorillas, can be found in Gene Plowden, *Gargantua: Circus Star of the Century* (Miami: E. A. Seemann Publishing, 1972).

21. Henderson, *Circus Doctor*, 214.

22. Surprisingly, no one has written a biography of Belle Benchley. She published two light books—*My Friends, the Apes* (Boston: Little, Brown and Company, 1942) and *My Life in a Man-Made Jungle* (Boston: Little, Brown and Company, 1943)—and a short article for an edited collection, *Wild in the City: The Best of Zoonooz* (San Diego, CA: Zoological Society of San Diego, 1985), 64–57. Margaret Poynter's *The Zoo Lady: Belle Benchley and the San Diego Zoo* (Minneapolis: Dillon Press, 1980) is a children's book.

23. C. R. Carpenter, "An Observational Study of Two Captive Mountain Gorillas," *Human Biology* 9 (1937): 193.

24. For Colo, see Nancy Row Pimm, *Colo's Story: the Life of One Grand Gorilla* (Columbus: Columbus Zoological Park Association, 2011), and Jeff Lyttle, *Gorillas in Our Midst: The Story of the Columbus Zoo Gorillas* (Columbus: Ohio State University Press, 1997).

25. Ernst M. Lang, *Goma, the Baby Gorilla*, trans. Edmund Fisher (London: Victor Gollancz, 1962), 37.

26. Ibid., 51.

27. Arthur J. Riopelle, "'Snowflake': The World's First White Gorilla," *National Geographic* 131 (1967): 44–48, and "Growing Up with Snowflake," *National Geographic* 138 (1970): 491–502.

28. Francine Patterson and Eugene Linden, *The Education of Koko* (New York: Holt, Rinehart and Winston, 1981).

29. Francine Patterson and Wendy Gordon, "The Case for the Personhood of Gorillas," in *The Great Ape Project*, ed. Paola Cavalieri and Peter Singer (New York: St. Martin's Press, 1993), 58–77.

30. Steven Pinker, *The Language Instinct* (New York: William Morrow and Company, 1994), 337–38.

31. Herbert S. Terrace, *Nim* (New York: Alfred A. Knopf, 1979), 7–22.

32. Ibid., 31.

33. Herbert S. Terrace et al., "Can an Ape Create a Sentence?," *Science* 206 (1979): 901.

34. Terrace, *Nim*, 226.

35. David Premack and Ann James Premack, *The Mind of an Ape* (New York: W. W. Norton & Company, 1983), 151.

36. Pinker, *The Language Instinct*, 339.

37. Noam Chomsky, "Are Those Apes Really Talking?," *Time*, March 10, 1980, 50, 57.

38. Noam Chomsky interviewed by Matt Aamis Cucchiaro, e-mail correspondence, 2007–2008.

39. Steven A. Anderson, *Doctor Dolittle's Delusion: Animals and the Uniqueness of Human Language* (New Haven, CT: Yale University Press, 2004), 275.

40. Sue Savage-Rumbaugh, Stuart G. Shanker, and Talbot J. Taylor, *Apes, Language, and the Human Mind* (New York: Oxford University Press, 1998).

41. Con Slobodchikoff, *Chasing Doctor Dolittle: Learning the Language of Animals* (New York: St. Martin's Press, 2012), 5.

42. Readers wishing to learn more about the complexities of the debate can consult Marc D. Hauser, Noam Chomsky, and W. Tecumseh Fitch, "The Faculty of Language: What Is It, Who Has It, and How Did It Evolve?," *Science* 298 (2002): 1569–79; Ray Jackendoff and Steven Pinker, "The Nature of the Language Faculty and Its Implications for Evolution of Language (Reply to Fitch, Hauser, and Chomsky)," *Cognition* 97 (2005): 211–25; and Tecumseh W. Fitch, Marc D. Hauser, and Noam Chomsky, "The Evolution of Language Faculty: Clarification and Implications," *Cognition* 97 (2005): 179–210.

43. Those interested in interspecies relationships, especially between humans and dogs, are referred to Donna Haraway, *When Species Meet* (Minneapolis: University of Minnesota Press, 2008). Despite excursions into philosophy that make for dense and not terribly informative reading, she has some interesting things to say. See also Kenneth Shapiro and Margo DeMello, "The State of Human-Animal Studies," *Society and Animals* 18 (2010): 1–17, and Tom Tyler and Manuela Rossini, eds., *Animal Encounters* (Leiden: Brill, 2009).

44. Patterson and Linden, *Education of Koko*, 80.

45. Anthony L. Rose, "Michael and Me," April 20, 2000, Bushmeat Project, http://biosynergy.org/bushmeat.

46. Note that this is the same name the Johnsons gave to one of the young gorillas sent to the San Diego Zoo. Given the notoriety of the movie, it's hard to understand why they did this and why the name stuck.

47. Additional early movies with gorillas can be found in Geoffrey Howard Bourne and Maury Cohen, *The Gentle Giants: The Gorilla Story* (New York: Putnam and Sons, 1975), chap. 2.

48. A sampling of the diverse literature can be found in Orville Goldner and George F. Turner, *The Making of King Kong* (New York: Ballantine Books, 1975); Ronald Gottesman and Henry Geduld, eds., *The Girl in the Hairy Paw: King Kong as Myth, Movie, and Monster* (New York: Avon Books, 1976); Cynthia Erb, *Tracking King Kong: A Hollywood Icon in World Culture*, 2nd ed. (Detroit: Wayne State University Press, 2009); Joshua D. Bellin, *Framing Monsters: Fantasy Film and Social Alienation* (Carbondale: University of Southern Illinois Press, 2005); Judith Mayne, "King Kong and the Ideology of Spectacle," *Quarterly Review of Film Studies* 1 (1976): 384–90; and Merrill Schleier, "The Empire State Building, Working Class Masculinity, and King Kong," *Mosaic* 41 (2008): 29–54.

49. Cited in Gottesman and Geduld, *The Girl in the Hairy Paw*, 12.

50. Cited in Ronald Haver, *David O. Selznick's Hollywood* (New York: Alfred A. Knopf, 1980), 77.

51. Erb, *Tracking King Kong*, 211.

52. *New Yorker*, July 27, 2011, 62, and May 28, 2012, 70.

Chapter Seven

The Future of the Gentle Giants

> And God blessed them, and God said unto them, Be fruitful, and multiply, and replenish the earth, and subdue it: and have dominion over the fish of the sea, and over the fowl of the air, and over every living thing that moveth upon the earth.
>
> —Genesis 1:28, King James Version

GORILLAS IN THEIR HOMELANDS

The issue of whether and how the "gentle giants" will survive needs to be examined from two perspectives. One relates to those still living in their tropical forest homelands. As should be clear by now, they face an array of human-induced pressures that either take the lives of individual gorillas directly or do so indirectly by altering required habitats in disadvantageous ways. To recapitulate, the most important of these include an ever-increasing encroachment of agriculture, the growing demand for tropical hardwoods, the mining of valuable minerals, the bush meat trade, civil wars, disease, and animal trafficking.[1] Trophy hunting is no longer a significant factor, and the role of climate change is uncertain, although it undoubtedly will have some impact via changes in vegetation cover. How the various factors play out in any given instance is contingent on geography, meaning where particular gorillas currently reside. Making location even more salient is the fact that gorillas cannot move elsewhere, mostly because people are increasingly everywhere around them.

Also geographical in expression are conservation efforts aimed both at protecting gorillas directly and at the habitats they need in order to survive. While many programs have come into being in recent decades, only a very small proportion of gorillas have been well served by them. Environmental, political, personnel, and financial factors all play roles in determining the effectiveness of conservation-related programs. The ways in which the interplay of pressures and conservation efforts work out can be illustrated by examining each of the four gorilla subspecies separately.

The isolated Cross River population is considered "critically endangered" because its scattered small troops cannot tolerate much in the way of further human interference. Yet people are pushing ever more deeply into the region in their search for agricultural land not already in production. As a result, fields and grazing lands replace forests, further fragmenting an already fragmented gorilla habitat. This is aided and abetted by a dry season when fires occur, and the more people there are in the mix, the greater the likelihood of fires being set either purposely or accidentally. Some logging is also going on, and hunting has picked up as well. According to custom, people living in the Cross River area do not eat gorillas and other primates, but commercial hunting has entered the picture. As is the case for most gorilla populations, killings and mutilations sometimes occur due to snares set to capture small animals. In the face of these pressures, the surviving gorilla groups, numbering perhaps as many as fifteen, have retreated into the more isolated rocky hills surrounding the Cross River.[2] At the moment, it seems as though enough forest corridors still exist to connect them, and a recent DNA analysis suggests that gene flow is occurring between different groupings.[3] This, however, might not be the case for much longer if present land conversion trends continue. Although no catastrophic disease outbreaks have been reported so far, the closer proximity of people could make human-to-gorilla disease transmissions an issue at any time. Tourists, however, are unlikely to be a cause for concern. No habituated groups exist or are likely to exist anytime soon (if ever), and finding free-ranging ones is almost impossible given their shyness and terrain difficulties.[4]

Mirroring the research discussed in chapter 5, sustained conservation efforts have just begun. A regional action plan was formulated in 2006, followed two years later by the creation of Takamanda National Park, a joint undertaking of the Cameroon government and partners like the World Wildlife Society. On the Nigerian side of the border, the earlier establishment of Cross River National Park failed to halt hunters and others from encroaching on gorilla territory.[5] It's too early to tell what this latest effort will yield.

An interesting attempt to broaden the base of support for conservation was a workshop for Cameroonian artists hosted by the Limbe Wildlife Centre, located in a pleasant setting near the large city of Douala. Thirteen artists attended during November 2007 and were provided the opportunity to view the gorillas at the Centre close up, including Nyango, the only Cross River one in captivity. Accounts agree about the workshop having been a success in producing high-quality works of art and in creating a greater sense of the need for gorilla conservation. As artist Emmanuel Tango remarked, "The Limbe workshop acted as a turning point in my artistic career. I felt some deep pain and emotion for those animals . . . I just asked myself: why should we have these fellow-beings facing extinction right before our eyes, and foreigners will travel thousands of miles to come and rescue this species when we sit back and watch?"[6] The larger question is whether there will be enough small steps like this one to make a difference when time is of the essence. According to some authorities, conservation efforts may be too late to save this isolated and still poorly known population from extinction.[7] Yet others, like Denis Ndeloh Etiendem, believe that it is not too late to save them if there is the will to do so.[8]

All of the threats impinge upon Western gorillas, the mix and severity depending upon where they are located. Of ever-increasing significance for those within moist hardwood forests is the intrusion of logging. Uncommon before 1980, the granting of concessions, mostly to companies in Europe, has become ever more common due to rapid tropical forest decline and a corresponding drop-off in hardwood timber coming from Southeast Asia. As roads are built, settlements follow. Their needs, along with commercial logging, shrink and fragment the area occupied by gorillas and make the likelihood of human contact with them more likely.

The recent expansion of palm plantations to meet the growing demand for palm oil is consuming its share of forested land. As with logging, the concessions are usually granted to international concerns, and while these may create jobs, they replace forest commons with private estates, thus denying people access to food, medicinal, and other resources. At the same time, of course, the space for wildlife, including gorillas, is reduced.[9] Unfortunately, detailed maps of both deforestation and gorilla numbers are lacking, and thus it is impossible to pinpoint the areas of greatest concern.

Throughout much of the region, and especially in Gabon, the Republic of the Congo, and the Democratic Republic of the Congo (DRC), gorillas have traditionally been hunted for food by some peoples. Adding to this source of killing is the bush meat trade, in which gorilla parts show up in town and city markets, even where outlawed.[10] When people are poor (and most are throughout this part of Africa), bush meat offers an opportunity to earn some cash, and the new roads have made the transport of meat much easier and cheaper. Workers in the logging camps also want their fair share. While it is true that great apes make up a small part of Africa's bush meat trade, maybe

4 percent at the most, the overall population impact is great, especially for gorillas.[11] This is because females give birth only every three to five years, and a long period of dependency follows. And we must not forget that in the wake of adults being killed for food, infants are often captured and offered for sale to interested buyers despite such sales being banned by international treaties. Markets for them continue to exist in parts of South America, the Gulf States, eastern Europe, and eastern Asia. A poignant and disturbing photo essay of the link between logging and bush meat can be found in *Consuming Nature*.[12]

Disease has suddenly become a major concern. From 1983 to 2000, an estimated 56 percent decline in Western gorillas and chimpanzees took place in Gabon and the Republic of the Congo from the combined impacts of hunting, logging, and Ebola hemorrhagic fever.[13] Ebola struck again in the Republic, claiming around 5,000 gorillas during an epidemic between October 2002 and January 2004.[14] The disease has been spreading, and 90 to 95 percent of infected gorillas die. It's highly unlikely that any transmission has come via humans. Indeed, the transmission so far has been the other way around, with humans contracting the virus from butchering gorillas infected by a natural intermediary yet to be identified.[15] It's possible that several species of fruit bats serve as reservoirs, although they may not be the actual source of the disease, and people do not eat them. These Ebola events are reminders that viruses and other common pathogens are more likely to be exchanged as contacts between *Homo sapiens* and other great apes intensify.[16]

Comparatively speaking, current numbers are on the side of the Western gorillas, but, as noted earlier, good censuses are few and trends unfavorable. And while national parks and reserves of various kinds have been demarcated in Cameroon, Equatorial Guinea, Gabon, the Central African Republic, the Republic of the Congo, and the DRC, they are, for the most part, what people call "paper parks," meaning that little or no effective control is exercised due to a lack of funds and qualified personnel. One that doesn't qualify for this negative epitaph is the Dazanga Sangha Special Reserve in southwestern Central African Republic. It has been well protected since being formed in 1990, and there's a habituated group that has drawn tourists willing to make the difficult trip.[17] Furthermore, probably somewhere in the neighborhood of 80 percent of Western gorillas live outside these areas. The populations with the best chances of sustaining themselves are located in swamp forests. Few people live within them, they harbor little in the way of current valuable resources, and they are mostly inaccessible, except by foot. This, of course, could change, as has happened in formerly remote forests occupied by orangutans. Due to isolation, little is known about swamp forest gorillas. There could be many more than thought, or there could be fewer.

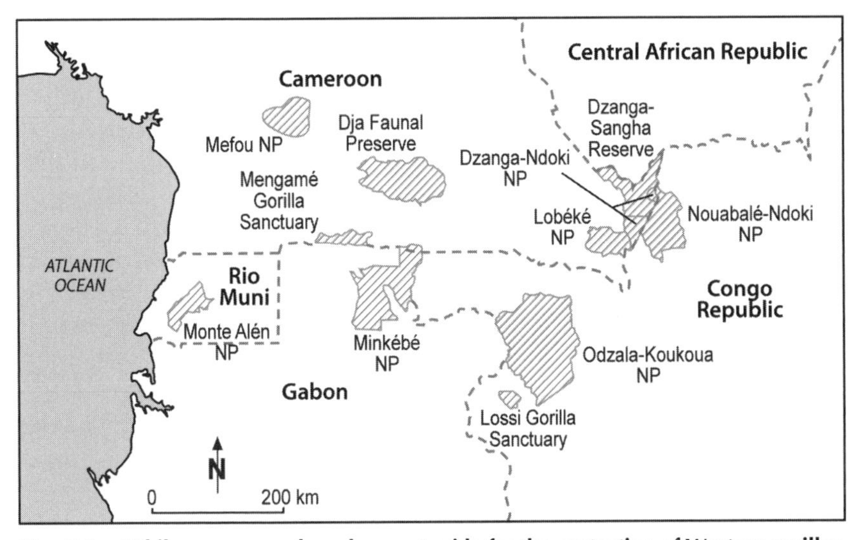

Map 7.1. While many areas have been set aside for the protection of Western gorillas, for the most part they exist on paper only.

Until recently, Eastern gorillas lived rather unperturbed lives because they, too, occupied rather remote areas with traditionally low numbers of people who, for the most part, refrained from eating apes. The earlier killing and capture by Europeans in search of Mountain gorillas seems to have had little if any demographic impact. But, as noted, the situation began to change dramatically in the 1990s as refugees poured into the area fleeing the civil conflicts affecting Rwanda and the DRC. They claimed large tracts of land, and for the first time gorillas became "meat" in a significant way. While the refugee crisis has abated, villagers from the densely populated lands to the east have been steadily drifting in, bringing with them land-consuming forms of slash-and-burn agriculture.[18] In addition, miners have come looking for coltan, gold, and diamonds. So far, logging hasn't been important due to the lack of a serviceable road network. Such a network is bound to be built, probably sooner rather than later.

The combination of human inroads has taken a particularly heavy toll on the gorillas in Kahuzi-Biega, with an estimated 50 percent loss occurring during the latter half of the 1990s. The situation has quieted some since then, and in 2007 park monitoring was upgraded with help from the Canadian Ape Alliance. A continuing presence during all these years has been former park ranger John Kahekwa. After the killing of the popular silverback Maheshe in 1993, he spent his own money to form the Pole Pole (it means "slowly") Foundation, with the intent of stemming the tide of decline by helping poachers find other means of sustenance, such as making wood carvings for tourists. It may be why a survey in 2010 counted 181 gorillas rather than none, and a few intrepid tourists have begun to trickle back. Research, though, has yet to resume.

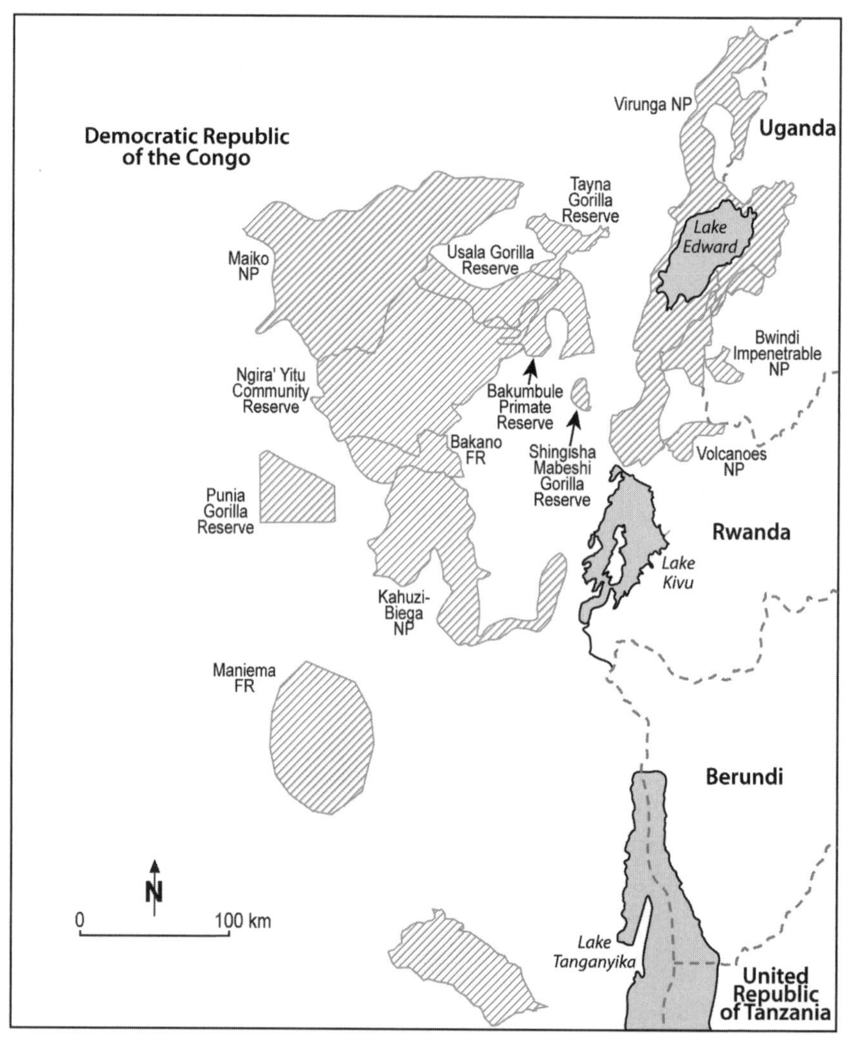

Map 7.2. The safe havens for Eastern gorillas have been seriously affected by civil strife.

Although several recent wildlife surveys have been undertaken in Maiko National Park, the status of its Eastern gorillas remains unknown. Hindering operations has been the presence of rebel groups dating back to the 1960s. Then, in the 1990s, members of the Interahamwe showed up, making life ever more dangerous for the rangers. In effect, there has been no control over the park for years. Similarly, the situations in the Itombwe, Punia, and Tayna reserves, along with the Shingisha Mabaeshi Forest Preserve, remain cloudy, although reports indicate that rangers have returned to Tayna after having been driven out. How gorillas are doing beyond these areas is anyone's guess.

Despite their few numbers, Mountain gorillas are not listed as "critically endangered" by the International Union for the Conservation of Nature. Indeed, as we've seen, their numbers are growing. Those in Rwanda's Volcanoes National Park are, at least for the moment, the most secure, living as they do in a rigorously policed area to keep out hunting and other unwanted activities, such as logging, mining, farming, and livestock herding. In addition, they are constantly under the eyes of researchers who look after their well-being, a conservation strategy second to none. The Mountain gorillas in Uganda's Mgahinga and Bwindi Impenetrable Forest parks, while less closely monitored, are also not under immediate serious threat. Although current growth trends are encouraging, caution must be exercised. These are isolated islandlike populations surrounded by densely settled poor human communities dependent upon agriculture and whatever else can be gleaned from the environment for sustenance. Disturbance and invasion, therefore, remain as possibilities.[19] Furthermore, such close proximity increases the chances of gorillas contracting diseases from humans and domestic animals. And as reports from Bwindi indicate, a problem exists with habituated groups wandering across park boundaries into human settlements.[20]

The most recent flashpoint is the DRC's Virunga National Park, where a new round of fighting between rebel groups and guards erupted in 2012, causing the country to close it to tourists. Their numbers had risen from 550 in 2009 to 3,300 in 2011, with further increases expected. The fate of the 200 or so Mountain gorillas residing within the park is unclear.

Gorilla trafficking remains an issue, as is illustrated by the "Taiping Four," consisting of a Western male, Oyin, and females Abbey, Izaan, and Tinu. The name comes from the Taiping Zoo in Malaysia, which purchased the gorilla from Nigeria's Ibadan Zoo in 2002 under the impression that they were captive-bred. An investigation by the International Primate Protection League found irregularities with the papers, and as it turned out these had been forged to cover up the fact that the four were taken from the wild as juveniles at an unknown point in Cameroon. This would have meant killing at least some if not all of their families. Officials of the Convention on International Trade in Endangered Species in Malaysia pressured the government to admit the error, which, after a delay, it did. Since the Taiping Four had passed through South Africa on their way to Malaysia, officials there offered to house them at the National Zoological Gardens in Pretoria, where they arrived in 2004. Soon thereafter, voices, especially those of Cameroonian authorities, began clamoring for the Taiping Four to be returned to their country of origin for placement within the Limbe Wildlife Centre. These efforts ultimately succeeded, and in November 2007 the Centre became the gorillas' new home. Since then, two have died, Oyin in June 2008 and Izaan in December of the same year. Abbey and Tinu remain residents of the Centre.

Limbe brings up another element of gorilla conservation: sanctuaries. Their primary objectives are the rescue and rehabilitation of orphaned, unwanted, and injured wildlife, hoping, where possible, to release them into appropriate environmental settings. Regarding apes, anthropologist Geza Teleki prefers "refugee" because the word "connotes acute plight caused by some destructive force and also because the traumas experienced . . . are equivalent physically and psychologically to the traumas experienced by displaced human refugees."[21] To accept his argument means expanding our sense of family to include all Hominidae.

In 2000, the Pan African Sanctuary Alliance came into being, with four major centers handling gorillas, Limbe, as noted above, being one of them. It began in 1993 as a cooperative venture of the Pandrillus Foundation, a nongovernmental organization initially focused on mandrills in Nigeria, and the government of Cameroon. The country's Ministry of Forests and Fauna is now in charge of operations. According to the latest information, sixteen gorillas are cared for, none of which seem destined for release.[22]

The John Aspinall Foundation, an offshoot of Howletts Zoo, is a registered charity supporting two Western gorilla sanctuaries in Africa, one in Gabon and the other in the Republic of Congo, operating under the name Projet de Protection des Gorilles. The zoo itself is also a sanctuary. The focus is on release, especially to help repopulate the Batéké Plateau in Gabon, from which gorillas had disappeared decades ago. The fourth sanctuary is run by another charity, the Cameroon Wildlife Aid Fund, which has close ties to the Bristol Zoo in the United Kingdom. It started out providing assistance to improve conditions for the resident primates at the Mvog Betsi Zoo in Yaoundé, Cameroon. More recently, the Fund has taken in orphans, including gorillas, of the bush meat and pet trades for placement at a site in Mefou National Park near Yaoundé, with rehabilitation and long-term care being top priorities. It has become a fairly popular tourist destination where gorillas and other apes can be seen up close.

Sanctuaries such as these are humane responses to the survival crises faced by many species. There are problems, however, especially with regard to gorillas. One involves numbers. How many can be kept at a given facility? And after capacity is reached, what happens when new refugees are found? Not all (and perhaps most) gorillas are suitable for release because of either physical or behavioral reasons. Then, too, once released, will they survive, meaning will they become members of functioning troops? The Aspinall Foundation has reported survival rates in excess of 80 percent, suggesting that the answer is "yes" for most. Less encouraging is what happened following the release of four young Mountain gorillas and one Eastern. All either died, disappeared, or had to be recaptured.[23]

Gorilla specialist Esteban Sarmiento would have wished otherwise, for, as he has remarked,

> ultimately, the most successful conservation measures should provide animals with independence from escalating human intervention, enable them to live out their lives by their own means, and promote self-sustaining populations. Raising animals in captive situations, or in exotic habitats in which they do not naturally occur, engenders human dependency and entrusts their survival to the whims of human economic and political concerns. Such rearing cannot be justified as conservation unless it is a prelude to reintroduction of otherwise extinct animals into their past natural habitats, and leads to a de-escalation of human dependency.[24]

This, however, could happen only in the best of all possible worlds, and for gorillas and other apes such is hardly the case. Consequently, we now turn to another home for them: zoos.

GORILLAS IN ZOOS

The origin of the word "zoo" can be traced to the formation of the Zoological Society of London in 1826, which two years later opened a collection of exotic animals in Regent's Park for purposes of scientific research and for a few selected people to view. Nonmembers had to be recommended by a Society member to gain entry. Menageries and other sorts of animal collections had, of course, existed since antiquity, and in the late eighteenth century the Jardin des Plantes in Paris added beautiful gardens to civilize the surroundings in which their charges were shown.[25] Other exhibits copied the pattern, and soon the use of Zoological Garden became common. In time, “zoo” took over as the more widely used designation, with the *Oxford English Language Dictionary* first recognizing the word in its 1867 edition. By the end of the nineteenth century, most large cities in Europe and North America had added such an institution to the urban landscape so that their citizens could gaze at wonders from faraway places in pleasant settings. Zoos became, in effect, “oases . . . secondary places of nature” for city dwellers, and they still are.[26] A number of factors coalesced to allow this to happen. The most important included increases in national wealth, democratization resulting from the growth of a middle class, exploration that brought back new exotic animals on a regular basis, and, on the science side, an obsession with the Linnaean classification of nature.[27] Many zoos would, in fact, organize their exhibits in such a manner.

In the early days, as we’ve seen, confinement in zoos meant a virtual death sentence for gorillas, as people knew next to nothing about their care and feeding and those that did survive for a while usually sat alone in stark cages. This is how architect and later zoo director David Hancocks described his first encounter with a gorilla at the London Zoo many years ago:

> I cannot recall what I was expecting from the zoo, though I remember being curious and eager. I was not, however, anticipating the shock of seeing a gorilla. It wasn’t his huge form that astonished me so much as the intelligence in his eyes. That, and the bitterly small size of his barren cage. This extraordinary animal, with his regal air, survived in a space no bigger than a garden shed. He was called Guy, and he sat on a concrete floor, soiled with his own excrement, looking out through bars and a glass window at a million people who shuffled past each year to gawk at him in his silent and solitary confinement.[28]

Much has changed since then, and here we’ll look at what has been done to improve gorilla well-being and longevity. To survey the whole of zoos would be a gargantuan task—fifty-two of the 224 zoos accredited by the Association of Zoos and Aquariums in the United States house gorillas, with many more in Europe—and thus I’ve decided to illustrate the general learning curve they have experienced by focusing on four case studies, Antwerp and Philadelphia representing older and smaller zoos and Columbus and Woodland Park newer and larger ones. Each has an interesting and somewhat different story to relate. Other zoos will be mentioned when appropriate. I won’t be dealing with safari parks and private zoos, which typically are not designed to keep gorillas because of their large size, high costs, and special needs. In addition, it would be difficult if not impossible to re-create their environments in the manner that safari parks have done for the African savanna and American prairie. Some background for each of the zoos will be provided to set the contexts in which gorillas have found themselves. In this discussion, I will not tell zoos what they should or should not do. That's for them and other experts to say.

ANTWERP ZOO

The Antwerp Zoo, or Zoo Antwerpen as it is officially known, came into being with the formation of La Société de Zoologie d'Anvers (soon to be La Société Royal) on July 21, 1843, making it one of the longest-running zoos in the world.[29] At the beginning the grounds, landscaped in the style of an "English Garden," were open only to the upper crust of society. On Easter Sunday 1862, the general public finally gained access. In the 1950s, the zoo acquired the Planckendael Wild Animal Park in nearby Muizen. No gorillas are kept there.

The grounds of the Antwerp Zoo are set within the city just beyond the main railway station and adjacent to a large public space, making it readily accessible to residents and visitors alike. But this also means that the zoo is locked into the roughly twenty-six acres currently occupied. Again, though, there is a plus side—compactness allows visitors to see the animals and enjoy the attractive landscaping in a leisurely half day. This is particularly advantageous for young children, who, like with most zoos, constitute a large share of visitors, whether with families or on school outings.

Many nineteenth-century zoos tried to emphasize the exotic, one tactic being the incorporation of structures based on Oriental themes. Antwerp is no exception, with its replica of an Egyptian temple on the Isle of Philae. Built in 1856 to house an array of African animals, the temple underwent extensive remodeling in 1988 and now features elephants and giraffes.

Little is known of the zoo's early history of keeping gorillas because all of the records pertaining to animals through World War II were either lost or destroyed. As previously noted, Ben Burbridge brought Quahalie there in December 1923, and from what can be determined she survived for just a year. Marzo arrived in 1925, and reportedly several people saw him late that summer, only to learn from a newspaper article of his death shortly thereafter.[30] Given Belgium's colonial rule of the Congo Free State, later the Belgian Congo, which lasted from 1885 to 1960, it's hard to imagine that other gorillas didn't make their ways to Antwerp at one time or another, especially since the city is a major European port of entry.

In any event, the zoo's modern history with gorillas begins with the arrival of a Western youngster named Gust on May 18, 1953, a gift from the governor-general of the Belgian Congo. He became a local legend, his name often used by Belgians for other gorillas, even after passing away on April 14, 1988.[31] In the late 1950s and early 1960s, Charles Cordier supplied the zoo with five gorillas, according to the best estimate. Females constituted the majority in the hope of starting a breeding program. A Western, Kora, became a companion for Gust. The peak year was 1968, with five Eastern and four Western versions included in the mix at the zoo. The 1950s also saw a substantial rise in the number of chimpanzees, and with orangutans also on hand, a new Ape House opened in 1958 to replace the sterile and stark cages that the inhabitants previously occupied.[32] An obsession with cleanliness led to such quarters, often referred to as the "bathroom" style, becoming the norm for most zoos immediately after the war. They were also inexpensive and conserved space. The new house included three outside and eight inside enclosures for the different apes, the outside ones oriented to catch as much sunlight as possible in Belgium's cool climate with notoriously cloudy skies. No bars obstructed the view by visitors. Terraces led down to a wide moat separating the apes from the public, an update of a common practice for nineteenth-century zoological gardens. Unfortunately, before the first year was out, a chimpanzee fell into the moat and drowned. The same thing had happened to Makoko, a silverback at the Bronx Zoo, in 1951. By way of response, officials in Antwerp placed a wire-mesh barrier in the shallows to keep the apes from reaching the deeper, more dangerous waters. The barrier showed its value early on when two playful young gorillas tumbled into the moat and then scrambled out on their own.

Polyglass lined the inside cages in order to give the apes protection from contact with human airborne diseases and to do away with such nuisances as people trying to feed and tease the animals, both of which, even today, remain zoo concerns just about everywhere. Antwerp may have been the first zoo to use the material instead of reinforced glass. Its use also gave visitors unobstructed close-up views of the apes. Other advantages included the ability to regulate temperatures inside and outside the enclosures and the containment of odors. While people love seeing animals, they often don't appreciate the smells associated with them. Quarters where the inhabitants could escape from view and make beds from hay existed inside. These also served as home during winter.

It was within such a setting that the first Eastern gorilla birth in captivity took place in 1968. Named Victoria, she is still very visible at the zoo. A second, Isabel, arrived in 1981. Sadly, she died prematurely in 1995.

The passing of the years led to the Ape House becoming obsolete, and consequently it underwent remodeling in 1989. The outside islands and moats were done away with, giving the apes more space behind new polyglass-lined enclosures. Previously, the house contained Western and Eastern gorillas, orangutans, chimpanzees, and bonobos, the first of which arrived in 1906, with others coming after World War II. As part of the remodeling, the bonobos were moved to Planckendael. By the time of its opening, Gust had died and his mate was sent away, leaving behind groups of chimpanzees, six Eastern gorillas, and three orangutans for the public to see. Virtually all zoos today have to make hard decisions about which animals to keep, and Antwerp eventually decided to forgo orangutans in order to give the gorillas and chimpanzees more space and attention. Furthermore, they fit within the zoo's long-standing Congo theme.

The new gorilla quarters emphasize environmental enrichment, with the larger main one having an array of climbing and swinging devices, along with various puzzles for the inhabitants to solve in order to receive rewards. The arrangements are changed regularly for purposes of mental stimulation. In addition, the gorillas have to find their food instead of handlers feeding them directly. There is also space for socialization to occur. By this time, zoos had come to realize that boredom is a serious health issue for gorillas. As before, the main exhibit area is connected to several smaller interior enclosures where the gorillas can rest and get out of public view when they so desire.

Time took its toll, leading the number of gorillas to decline from six Eastern ones in 1989 to three in total—Victoria and another Eastern female (Amahoro) and a Western silverback (Kumba)—by the middle of the last decade. Sensing the need to increase the size of the group for social reasons, an adult Western female, Mambele, joined the others in 2011. No pool of captive Eastern gorillas exists to draw from. The introduction proceeded cautiously, and after some early shows of aggressive and agnostic behavior, the four adapted to the new situation in little more than a week. Since acceptance by a silverback is crucial to successful group integration, the bond formed by Kumba and Mambele came as good news. While this took place, Victoria and Amahoro formed a closer relationship. As of now, all seems peaceful.[33]

Gorillas are one of the animals included in a worldwide captive breeding program, and only the Western version is involved. A detailed stud book for them is kept in order to control numbers and to ensure adequate genetic diversity among newborns. This means that zoos do not own gorillas as they once did. Instead, the broader zoo world owns them for reproductive purposes. A part of the policy is to avoid hybrids, and consequently the Antwerp population is a non-breeding one.

Spearheaded by the New York Zoological Society, education, conservation, and research have become watchwords for most zoos. With regard to the general public, education and conservation are usually combined, the former seeking to create a broader base of support for the latter.[34] How this is done varies. Signage per se is not thought to be overly effective, as people come to see animals, not read about them. Although animal demonstrations have become a zoo norm, gorillas are not very well suited to this educational approach. According to Angus Martin's observations in Australia, the most important education is likely to be informal, such as seeing gorillas in a re-created forest environment and thereby making the connection of the need to preserve such habitats for them[35] or, as at Antwerp, to see them active and socializing, thus emphasizing the connection to humans. In response to modern trends, the Antwerp Zoo has developed a video game in which points are given for correctly identifying the creator of a picture: a human or a gorilla. According to officials, the game is popular with children and adults alike, although it is not played very often in a competitive manner. The educative value has yet to be determined.

Virtually from the beginning, Antwerp has placed an emphasis on research, which today is carried out by its Centre for Research and Conservation, supported financially by the government of Flanders. It also has a close working relationship with the University of Antwerp, making it one of the few zoos to have such an arrangement. Because Antwerp is a comparatively small zoo, it is unable to conduct the far-ranging research carried out by some larger ones like the Bronx Zoo and Zoo Atlanta. Thus, as far as apes are concerned, the focus is on the bonobos at Planckendael. Gorillas, however, have not been forgotten. This is illustrated by the work of Denis Ndeloh Etiendem and his colleagues in the Centre's field site in Cameroon dealing with Cross River gorillas. And, as noted earlier, gorillas are viewed as an important part of the Congo theme that Antwerp seeks to maintain. Furthermore, like chimpanzees and bonobos, the gorillas serve as ambassadors for other, less compelling species and for their compatriots living outside the bounds of zoos. Despite being relatively inactive compared to the other two African great apes, gorillas remain at the top of the public popularity list, thus serving as "exhibition value" for the Antwerp Zoo, meaning that they help draw visitors.[36] Size probably plays some role in this, but, similar to Schaller and Hancocks, I think the eyes may matter most. Take a look sometime to see what you think. But be subtle about it, as some gorillas are wary of direct eye contact. And try to be lower than they are when you look. It's less threatening.

Another new Ape House is in the offing, and space limitations will prohibit building anything resembling the naturalistic forest environments pioneered by Seattle's Woodland Park Zoo described later.[37] No matter the eventual design, researcher Jeroen Stevens sees a need to continue improving environmental enrichment as a top gorilla priority. Although diets are now better attuned to their needs than was the case in the old days, he believes that more has to be learned about nutritional requirements. The same holds, he feels, for invasive medical procedures. The search thus goes on to provide a better life for the Antwerp Zoo's gorilla charges while at the same time improving the experiences of the million or so visitors each year in ways that will broaden the base of support for species and environmental conservation.

PHILADELPHIA ZOO

Philadelphia is home to the oldest—they prefer to say the first—zoo in the United States. To make the point, its publication is titled *America's First Zoo*. The story begins on March 21, 1859, when the Commonwealth of Pennsylvania formally chartered the Zoological Society of Philadelphia and gave it the right to the "purchase and display of living wild and other animals."[38] The gates, however, did not open to the public until July 1, 1874, due to the American Civil War and financial difficulties. A need also existed to find a substantial plot of land, which eventually was secured at the south end of Fairmont Park, the largest municipal park in the country. As is generally the case, the zoo sought to combine both entertainment and education, and to encourage people to visit, the admission fee was set at a quarter for adults and a dime for children. It would stay the same for half a century. Still, after a fast start, visitations declined. Even so, the number of animals continued to grow from just under 900 at opening, and attempts were made to give them more freedom of movement beyond their cages. Then came the several economic downturns persisting into the 1890s. Despite its own budgetary problems, the city of Philadelphia saw the zoo as an asset and thus came to the rescue on several occasions in order to keep it open.

On the animal side of the equation, zoo officials became increasingly concerned about health, and to find out the causes of death, they enlisted University of Pennsylvania pathologists Charles B. Penrose and Cortland Y. White in 1901 to conduct regular detailed necropsies, a first for zoos. In 1906, they discovered that tuberculosis ran rampant among primates, and to stem the tide, work began on installing iron partitions between the cages. Unfortunately, some of the workmen turned out to be carriers, and within a short time the disease claimed the lives of half the monkeys. With health now at the forefront of issues, Dr. Herbert Fox, another pathologist, joined the staff that same year. Still, tuberculosis reappeared from time to time, and in 1930 glass panels were added to the cages in order to prevent transmission from the public. This, along with regular testing, ultimately put an end to serious outbreaks.

An additional important health development came from research conducted by Dr. Ellen P. Corsen White from 1918 to 1928. She had been studying bone diseases among poor inner-city children and came to the zoo to see if primates suffered from similar afflictions. A horrific situation greeted her eyes, one of early death and severe bone deformities, especially among *Cibidae* monkeys. To solve it, she began research on developing a supplement that could be added to the diets of most of the zoo's animals. Those in charge, however, didn't buy the idea. Convincing them fell to Dr. Herbert L. Ratcliffe, who named the product Zoo Cake and oversaw its implementation as part of the daily feeding routine in 1935. With only minor changes, Zoo Cake remained in use until being replaced in 1989 by new commercially available biscuits that cost less and stored better. Zoo Cake, though, had opponents, the most vocal having been the German biologist and zoo director Heini Hediger, who repeatedly challenged Ratcliffe's claims for the health benefits it brought. Indeed, to him, Zoo Cake had just the opposite effect. Controversy aside, Zoo Cake served to make improved diet a top priority at all quality zoos, and several others did adopt it.

From a preventive health standpoint, Bamboo and Massa entered a zoo that would have been hard to beat at the time. Furthermore, when a necropsy on Bamboo, whose top weight reached 435 pounds, revealed a serious heart condition, health officials responded by keeping Massa at a trim 300 pounds, a factor that likely contributed to his longevity. Each day he ate two pounds of Zoo Cake, along with carrots, oranges, kale, and cabbage. Still, Massa needed more help from the zoo because at age thirty-eight, and now the oldest gorilla in captivity, he had developed a severe case of chronic sinusitis. The examination required administering a potent anesthetic, and the surgery team discovered that an infection had eroded his nasal bones. With these patched up, he went on penicillin to control the infection and quickly came back to form. But another problem surfaced: serious dental issues, including loose teeth, which could very well have meant his demise. Massa hadn't responded well to the previous anesthetic, so this time they decided to go with a new, milder version. A second dose put him under, and seventeen teeth, including an upper canine, were extracted, completing the first dental surgery on a gorilla ever. Once again, Massa quickly recovered and learned how to chew food with his front teeth.[39]

Bamboo II joined Massa in 1961 but didn't thrive to the same extent and died six years later. The big push to add gorillas came in 1969–1970 with six arrivals, all wild caught, in an attempt to establish a breeding program. They were housed together in the Rare Mammal House, built in 1965, and later moved to a new World of Primates building. The first pregnancy among them in 1975 resulted in a stillbirth, but five years later Jessica arrived on the scene, followed by Justin, Chaka, Anaka, Kola, Uhuru, Tufani, and Maandazi from 1982 to 1995.

Fearing that the trauma would be too much for him to bear, Massa remained in his old home, its metal bars replaced by clear plastic in 1975. On June 2, 1983, he was transferred to the Rare Mammal House, there to live out his final years.

Plate 7.1. Statue of Massa standing guard at the Philadelphia Zoo. *Source:* Reproduced courtesy of the Philadelphia Zoo.

Tragedy struck shortly after midnight on December 24, 1995, when a fire started in the World of Primates building. Two security guards had smelled smoke, but they thought it came from nearby railroad tracks, a common occurrence. No one heard the fire alarms, which were in working order, and the building did not have a sprinkler system. Later attributed to an electrical malfunction, smoke, not flames, took the lives of six gorillas, including a newborn, along with three orangutans, four gibbons, and ten lemurs. After recovering from the shock and grief, zoo officials went about the task of rebuilding, and in 1999, with funding help from a variety of sources, the PECO Primate Reserve came into being, housing eleven species. Among them were four gorillas, three obtained from other zoos, and Chaka, a member of the fire-stricken group, who had been on loan at the time for breeding purposes. Outside is a bronze sculpture of Massa, created in 1980 to celebrate his fiftieth birthday.

Health status monitoring by professional veterinarians is the norm today among certified zoos. Complete physical exams, radiographs, serologic testing, fecal analyses, and vaccinations are all part of preventive programs. Similar to many others, the Philadelphia Zoo uses positive reinforcement techniques to teach gorillas (as well as other primates) behaviors that help to achieve better health care, including the presentation of relevant body parts for examinations and injections.

Like Antwerp, the Philadelphia Zoo is confined to a limited space of forty-two acres. Consequently, also like Antwerp, creativity has had to be used to make sure that the six gorillas currently resident in the PECO Primate Reserve have environments that provide for social interaction, intellectual stimulation, and privacy in somewhat confined quarters. And although orangutans are where research and conservation efforts are currently focused, gorillas remain a high priority, as exemplified by the zoo's joining a national effort to create bachelor groups in order to replicate a naturally occurring social arrangement. So far, the two young males involved seem to be getting along quite well.

The Philadelphia Zoo's future is spelled out in its Strategic Plan 2012–2017, which, "by connecting people with wildlife," seeks to create "joyful discovery" and inspire "action for animals and habitats." In other words, pleasurable experiences are seen as motivating concern for wildlife and consequently a commitment to conservation. An innovative aspect of the plan is the construction of elevated trails across the zoo, designed in such a way as to give gorillas and other inhabitants an opportunity to enrich their lives by traveling to and exploring territory beyond their individual quarters.

COLUMBUS ZOO

Today's Columbus Zoo and Aquarium began as the Columbus Zoological Park in 1927 with just a few donated animals, including seven reindeer, in its possession.[40] Growth proceeded slowly in the 1930s and 1940s, and the city took over operations from the Zoological Society in 1951. Conditions, of course, changed in the 1950s after Colo appeared on the scene. But the glow she provided soon dimmed as quantity overtook quality, and by the mid-1960s the zoo received a "poor" rating due to too many animals crowded into too little space. On top of this, the four gorillas and two orangutans in the population were diagnosed with a human strain of tuberculosis, leading many to think their days were numbered, as exemplified by a *Columbus Dispatch* story carrying the headline "Colo Death Predicted in 90 Days." Instead of bowing to recommendations to put the apes down, the staff opted to begin a course of drug therapy, and within a short time all six patients were on the road to recovery. Following the international trend, glass partitions were installed to help reduce the likelihood of human airborne diseases reaching the apes. Little else, though, improved life for the gorillas. Viewed as too dangerous to keep together, they inhabited separate concrete and steel cages, which fully exposed them to public view. Their activities consisted mainly of eating, sleeping, playing with a toy or two, and staring back at the crowds.

So bad, in general, had conditions become that some influential community members suggested closing the zoo. A successful bond referendum to improve basic infrastructure forestalled that eventuality, but a 1976 plan for a $4 million habitat to house twenty gorillas went begging when not enough money could be raised privately.

In 1978, Jack Hanna took over as zoo director. Full of passion and energy, he would institute wide-ranging changes by launching a "New Columbus Zoo" campaign, and in time he became the face and voice of zoo animals not only in Columbus but throughout North America and even beyond. He's a regular on television to this day. One of Hanna's first acts was to raise money for an outdoor gorilla environment where the animals could exercise in open air when weather conditions permitted and where people could see them romping around in playful moods, not just sitting in a cage.

Two other people helped keep the momentum headed in a positive direction. In 1983, Dian Fossey paid a visit. She emphasized the need to replicate natural conditions as far as possible. Straw should be put on cage floors so that the gorillas could build nests, and they should forage for food in the straw instead of being fed directly. Diet itself had to change. In general, the gorillas were still being fed some meat, whereas Fossey recommended that they should be getting primarily fruits and vegetables, with supplements like yogurt and hard-boiled eggs for protein. This shift, she said, would lead to weight loss and more energy, changes that eventually happened. In addition, mothers needed to care for their newborns if at all possible rather than hand raising them, as practiced at Columbus and most other zoos fortunate enough to have births.

The biggest voice promoting change in how zoos exhibited their animals belonged to John Aspinall, who ran Howletts Zoo (now called Howletts Wild Animal Park).[41] A shady figure, he had made a fortune from gambling and clearly preferred animals to people. After winning big in 1956, Aspinall purchased an estate near Canterbury, Kent, England, and began fulfilling his animal dreams. The centerpiece involved the construction of open spaces designed to house larger, more diversified social groups, which for gorillas meant having ones beyond their reproductive years. These are prominent members of healthy, functioning families in nature. He also pioneered the use of extra-thick straw bedding and built the first devices for physical and mental enrichment. The changes led to more normal behavior for his gorillas, and along with it came more successful conceptions and births, ones in which most of the mothers, not staff, took care of their newborns.

These successes caught Jack Hanna's attention, and when the Columbus Zoo received $700,000 in 1984 to build new quarters, he used the Aspinall model to construct "Gorilla Villa," a large, round, multistoried outdoor earthen playground where the apes could climb, swing on ropes, and socialize. That same year, Hanna acquired a male, Mumbah, from Howletts and added Lulu to create a hoped-for breeding couple.

The opportunity to break the pattern of hand raising infants came with Jumoke's pregnancy. She had never seen a live birth or witnessed a mother caring for an infant. The first step involved getting Jumoke to trust her handlers. Without this, nothing else would be possible. Once established, they used a small stuffed gorilla toy to show Jumoke proper ways to hold an infant. Hearing motherlike vocalizations also became part of her regular routine, and she learned to tolerate some procedures, such as an ultrasound probe that would be used during pregnancy. Early in the afternoon of January 26, 1997, Jumoke gave birth to Jontu and immediately began hoped-for care, and she was calm enough to allow her handlers to touch him from the start.

A problem soon surfaced. The father, Anakka, young and aggressive by nature, showed no signs that he knew how to behave around an infant, and Jumoke became agitated when he entered the cage. To forestall a violent confrontation that could result in serious injury or even death, he was allowed only short, supervised visits, and zoo officials concluded that no one model could be used for births. How to respond would depend on individual circumstances. Anakka proved the wisdom of this approach, and today he heads Group I, the largest, with five under his leadership.

An important circumstance involved what to do when a mother refused to care for her newborn. The learning process began in 1986 by closely monitoring Fossey, the first mother-tended gorilla infant at the zoo. This led to improvements in hand-rearing practices and the decision to form a diversified age/sex group built around a program "where gorillas would be raised as gorillas and not humans." A detailed "protocol" was subsequently developed. The next step involved trying to find out if a surrogate mother would solve the problem. Colo was chosen to be the pioneer despite not having raised her own three offspring. She had, however, helped mother Jontu. Colo readily took to the task, and Columbus Zoo officials decided to establish a Surrogate Program that soon had other zoos sending their young for care and social bonding. It currently works as follows:

> A keeper will be responsive to the infant's needs twenty four hours a day, seven days a week, including at night. . . . The infant is never left alone. Keeper and infant spend their day in a room beside other gorillas. The infant has tactile, visual, auditory and olfactory stimulus with gorillas and the potential surrogate. Keepers simulate appropriate gorilla vocalizations, feeding and play behavior, as they are important components of this program. Training the infant gorilla to come to the mesh for the bottle is important. He/she must learn to receive a bottle through the mesh for nourishment, as the infant needs to be fed once in with gorillas. The infant has the opportunity to become familiar with his environment by playing and sleeping in the hay, going through the chute system, transferring through doors and spending time in the outdoor area. This process makes the transition to the surrogate stress free.[42]

The keeper helps select the most appropriate surrogate mother by carefully watching reactions to the infant. Usually after three to five months, the two are introduced, and one to two months later, they join the entire group. So far, thirteen infants, including six from other zoos, have been raised in the Surrogate Program. In addition, the Columbus Zoo distributes surrogate packages, including ones for bonobos and orangutans, as well as to gorillas. Those for chimpanzees and gibbons and siamangs are in the works.

Plate 7.2. The gorilla habitat at the Columbus Zoo has the various devices used to keep gorillas physically and mentally active. *Source:* Reproduced courtesy of the Columbus Zoo and Aquarium.

Under Hanna's leadership, the Columbus Zoo started its climb upward, eventually to reach a position of world respect. In 1989, it purchased an adjacent golf course and added 130 more acres in 1999, the two acquisitions doubling the original size and making the zoo the largest municipally run one in the country. Hanna stepped down in 1995, becoming director emeritus, a position he still holds. Before leaving, a 10 Year Master Plan had been drawn up, with the zoo grounds to be reorganized around biogeographic regions. Phase I of the African Forest exhibit was completed in 1998, followed two years later by phase II, the Congo Expedition. Within it is a large indoor space known as "Gorillas in the Round," where visitors can get a 360-degree view of the ones who happen to be inside at the time.

By the beginning of 2012, the gorilla population had grown to sixteen in three groups.[43] On May 18, the number declined by one when Mumbah passed from a heart condition. Lulu preceded him in January 2011. While Mumbah sired only one successful pregnancy during his many years at the zoo, he emerged as the patriarch of the Surrogate Program, playing the role sixteen times overall. Colo served as a surrogate mother for three more infants but is no longer part of a group, preferring instead to be alone much of the time. She deserves to have it her way.

In step with other major zoos, Columbus has launched a variety of educational and conservation programs. Among the latter is ongoing support for the Dian Fossey Gorilla Fund, the Mountain Gorilla Veterinary Project, the Mbeli Bai Study of Western Gorillas, the Bushmeat Crisis Task Force, and, more recently, Conservation of the Cross River Gorilla.[44]

WOODLAND PARK ZOO

In 1899, the city of Seattle took possession of an estate just outside of town called Woodland Park.[45] It had been the property of Guy Phinney, who had made a fortune from lumber and real estate, but he died early in life while creating an English garden–like atmosphere on the grounds. Among other things, it contained formal gardens and a small herd of deer. In 1903, a private zoo in the city went out of business, its animals then being transferred to the park facility. Shortly afterward, the city contracted with the Olmstead Brothers firm, which had helped design Central Park in New York City. The finished facility maintained the formal gardens and added spaces for animals on the edge of the park. During the next several decades, the number of exhibits increased on a fairly regular basis, with a Primate House (although without apes) added in 1911. Over considerable public opposition, the construction of a six-lane highway in the 1930s bisected the original park space, confining the zoo to the cleared "Upper" grounds, whereas the "Lower" grounds remained forested. Although the Depression imposed financial hardships on the city and the zoo, WPA projects added several new facilities, and by the end of World War II, slightly over 200 species called Woodland Park home.

In 1953, Bobo became Woodland Park's first gorilla, and in 1956, Fifi, also a wild-born Western gorilla, joined him as a prospective "bride." Both went into a Great Ape House completed in 1957 with the by now usual glass panels protecting the inhabitants. This arrangement didn't work out according to plan, as Bobo refused to accept Fifi's presence. He died in 1968, and a postmortem revealed Klinefelter's syndrome, a chromosomal disorder associated with hypogonadism and lessened fertility.

In 1965, the Seattle Zoological Society (now the Woodland Park Zoological Society) came into existence. Three years later, the zoo received three young gorillas from an unidentified source. Why the donor has never been revealed isn't clear. In any event, only one, Nina, lived to maturity. Two more donated infants arrived in 1969, a female, Kiki, and a male, Pierrot, quickly renamed Pete. In 1971, Kiki left for the Honolulu Zoo, living there until her death in 1978. Back at Woodland Park, Pete and Nina formed a couple, and in 1976 the zoo welcomed the arrival of its first baby gorilla, Wanto. Nina eventually presented Woodland Park with three more little ones.

The 1960s and early 1970s saw the zoo stagnate to the point where it had become a civic embarrassment. This certainly applied to the gorillas, which sat forlornly in stark cages or, at the other extreme, displayed signs of aggressive behavior. It wasn't uncommon for spectators to poke fun at them by whooping, banging on the glass, and jumping up and down. Others felt saddened to see the apes and others in such deplorable conditions. Possible closure of the zoo hung in the balance if change didn't happen soon. The change turned out to be a bold act; David Hancocks was put in charge of creating completely new exhibits based on environmental themes. A prototype existed in the African Plains exhibit at the Bronx Zoo, which went on display in 1941. Woodland Park was now to be completely redesigned around similar kinds of ecological zones and with emphasis put on the animals. The comforts of visitors would have to take second place. Work began in 1976, and in the same year Hancocks took over as zoo director.

For the gorillas, this meant home becoming the Gorilla Forest, an outdoor exhibit of varied terrain and live vegetation consisting of trees and wild-appearing grass designed to create a sense of "landscape immersion."[46] Objections were voiced almost immediately. The gorillas would destroy the vegetation, the public wouldn't be able to see the animals clearly, and the seemingly unkempt vegetation would prove to be an eyesore. In early summer, the experiment began, with Kiki selected to be the pioneer. After taking a few tentative steps, he accepted his new surroundings, as did the rest of the family within a week. They did destroy some vegetation but nothing that couldn't easily be replaced. One day, though, a problem occurred: Kiki managed to get out. Fortunately, neither he nor anyone else suffered harm before handlers returned him to the forest. To make sure that such an incident didn't occur again, an electrified fence was installed on the inside of the dry moat separating the exhibit from the walkway.

According to Hancocks, the change has been rewarding for all. As for the gorillas,

> Ritualistic threats between the two adult males were muted. Sudden body rushes were no longer resolved by physical contact but by one of the animals simply moving away and out of sight. Extreme inertia gave way to exploration and play. The group was undeniably more relaxed. They appeared content.[47]

Plate 7.3. A silverback in a simulated tropical forest setting at the Bronx Zoo. *Source:* Author photo.

Furthermore, "Zoo visitors who once stood in the grimy corridor of the old ape house, passively gawking at or mocking the animals with whoops and shuffling jumps, now stood in small clearings amid dense vegetation and did not shout or howl or, often even talk, but occasionally whispered to each other with wonder in their eyes."[48] These kinds of results prompted the San Diego Zoo to move in the same direction, as did Zoo Atlanta, which now has the largest landscape immersion gorilla exhibit.

The Woodland Park Zoo currently has three gorilla groups with a total population of ten.[49] Group 1 consists of two originals, Pete and Nina. He serves as the patriarch for all the others although he is no longer able to reproduce because of age and a kidney disorder. Group 2's silverback is VIP, who came via the Netherlands and Boston's Franklin Park Zoo in 1996 and has since sired five daughters. His family consists of Jumoke; wild-born Amanda, obtained from the Toronto Zoo in 1999; Calaya, an offspring of VIP and Amanda; and Uzuuma, also born of VIP and Amanda. Leonel is the leader of group 3. He arrived in 2008 after having spent time at several other zoos. Two females make up the rest of his group, Nadiri and Akenji, both of which Jumoke refused to mother, leaving the task to Nina.

In 2004, Woodland Park released its "Long Range Physical Plan." Although calling for many changes to the grounds, the Gorilla Forest remains as before. Indeed, the focus is on upgrading the experiences of visitors through "fostering public understanding of the history of animal life and its relationship to ecological systems." Selected species will be chosen for "care and propagation" in order to enhance "public awareness of human impact on animals and their environments." Gorillas are not mentioned specifically, but it's certain they will be one of the "selected species" given their endangered status and importance to and history of the Woodland Park Zoo.

Despite all the progress made in keeping and exhibiting their gorillas, with future plans indicating that more improvements will be forthcoming, zoos remain the object of criticism, some of it highly vocal. Regular contributors are the Society for the Prevention of Cruelty to Animals (SPCA), People for the Ethical Treatment of Animals (PETA), and the Humane Society of the United States, the last of which recently absorbed another animal rights organization, The Ark Trust. The basic charge is the same as it has been since zoos were formed, namely, that the animals are prisoners, kept only for human purposes, whether for entertainment, education, and conservation or as objects of study. To many of this mind-set, zoos should be shut down, and at the extreme end there have been calls for and even attempts at setting the animals free. This, of course, would mean death for most, whether by exposure or by killing. The latter is exactly what happened to the lions and Bengal tigers released from a private animal reserve just outside Zanesville, Ohio, on October 19,

2011. What if gorillas were to be released? Immobilization likely would be the first option, and, if successful, they would find themselves back whence they came. But if people and property were put at risk, then death might be their fate—a no-win scenario in either case.

The Great Ape Project (GAP) poses a different kind of challenge. It stems directly from Australia-born Peter Singer's 1975 book *Animal Liberation*, which puts forth an argument for animal rights based on moral philosophical principles, in essence saying that all sentient life, not just *Homo sapiens*, deserves the same kinds of respect and protections. Because of their biological closeness to us, great apes were chosen to begin a movement designed to create a broader "community of equals" in a manner similar to Geza Taleki. To this end, the GAP formulated "A Declaration on Great Apes" centered around the principles of "The Right to Life," "The Protection of Individual Liberty," and "The Prohibition of Torture," with each "to be enforceable at law."[50] No call is made for the release of the great apes from zoos since, as Singer notes, they, like many others under human care, would not be able to survive on their own.[51]

As you can imagine, GAP has sparked considerable debate. There are those who, mainly for religious reasons, support the specialness of humans. Others argue that GAP is out of touch with the realities surrounding great ape conservation, both within and outside of zoos, while still others think that singling out the great apes for special attention because of their closeness to us actually increases speciation.[52] GAP remains active, with its international headquarters located in Brazil, where the local one has established sanctuaries for chimpanzees released from research facilities. The size and strength of gorillas have kept them from being used for research purposes of the biomedical kind experienced by chimpanzees. Beyond this, it seems that GAP will remain primarily a theoretical voice for animal rights, at least for the foreseeable future.

Sometimes zoos are the focus of local criticism, especially when a charismatic inhabitant is injured or dies. A good example is Tiny, who in October 2010 became the first gorilla to be born at the London Zoo in twenty-two years. Naturally enough, the birth generated much enthusiasm around the country, and afterward Tiny's every development was closely followed. His first venture to explore the zoo's Gorilla Kingdom all by himself became a very big story in the newspapers. Unfortunately, Tiny's father Yehboah died from diabetes shortly thereafter. As a replacement, London Zoo officials brought in Kesho from the Dublin Zoo and slowly began introducing him to the females, the approach having been sanctioned by the European Association of Zoos and Aquaria. All went well, as did the first meeting between Kesho and Tiny. However, on the second attempt, a group scuffle broke out during which Tiny suffered a broken arm. The keepers rushed him to surgery for repair, but afterward he couldn't breathe by himself, and he died on May 31, 2011.

Outrage surfaced immediately, the London Zoo being charged with incompetence, among other things. Such events do occur in the wild. George Schaller, for example, found a mortality rate of 40 to 50 percent during the first six years of life of Mountain gorillas, the highest rate occurring in the first year due to disease and injury.[53] Zoo officials readily admitted the dangers of introducing a new silverback when the group contains the young of another male. This had to be weighed, they later said, against leaving the females without a male, something considered unnatural and detrimental for them socially. The critics were hardly mollified. Why, then, wasn't greater care taken, maybe delaying the introduction of a strange male or, better yet, not introducing him at all? The London Zoological Society clearly suffered a black eye from the incident and undoubtedly lost some support. Still, incidents like this one soon die down as other news captures the headlines. As for Kesho, he suffered his fair share of vilification and eventually wound up at the Longleat Safari Park in Wiltshire, there to rejoin his younger brother Alf.

The fact of the matter is that metropolitan zoos are here to stay. They have become an institutionalized part of modern life into which careers, time, and money have been invested. In addition, since their inception, entertainment has been at the heart of the zoo experience, especially of the family-related kind. There's more and more competition today, and zoos are trying to keep pace. Thus, we see pleasant food courts becoming the norm and people dressed as animals parading about for purposes of fun. In addition, special events, such as "Brew at the Zoo" and "Trail of the Lorax," are designed to bring in customers who otherwise might not come. While some people find this objectionable, there's nothing wrong with entertainment, especially if something else of value accompanies it without harming the animals. Right now I think that the "something else" has to do with the increasing separation from the world outside that characterizes modern living. While the trend has been in progress for a long time, momentum has picked up recently as people have become more and more attached to their electronic gadgets and indoor life. Indeed, even when outside, many people's minds seem to be inside. A zoo is one place where wildlife can be readily seen in the flesh, and some visitors may come away enriched from the experience through developing a better appreciation of life beyond one's own, including its importance to us in many different ways.[54]

What should be shut down are backyard private zoos and game parks. These are usually the products of someone's fantasies and/or desire to make money from admissions fees. As the owner of a recently opened Wild Animal

Park in Chittenango, New York, stated, he's living his "dream," and while it's a "business," he considers the animals "family."[55] Similar to laboratory research, the nature of gorillas comes to their rescue. They can't be easily kept at such facilities. Furthermore, being classified as "endangered" means that gorillas are carefully monitored and that, as the Taiping Four case illustrates, an illegal purchase would be uncovered in short order.

So what does the future hold for gorillas? Will they be around to share the planet with us, or are their days numbered? I think the best way to get close to an answer is to broaden our perspective in order to take into account where the world more generally seems to be headed. Some analysts have proclaimed the current era the Anthropocene, meaning, either directly or indirectly, that humans, not geological or other forces of nature, are in control of what's happening to the earth. And, if you think about it, what we now call natural places are such only because we afford them certain protections from becoming thoroughly humanized or, in the stronger word of Henry Fairfield Osborn Jr. made more than six decades ago, "plundered."[56] In a sense, the charge to have "dominion over the fish of the sea, and over the fowl of the air, and over every living thing that moveth upon the earth" is being fulfilled. Accompanying this "dominion" is a dramatic increase in the rate of species extinction to the point where a sixth major period may be under way. The last one occurred about 65 million years ago at the end of the Cretaceous period, when in the neighborhood of 75 percent of species, including dinosaurs, disappeared.

Overall, barring a major unforeseen catastrophe, gorillas are not likely to join the ranks of the extinct anytime soon despite our disruptions of their habitats and other insults. As discussed above, Western gorillas still survive in portions of Africa's lowland equatorial rain forests, and enough of them exist in North American and European zoos to provide for future reproduction now that zoos know how to do this. In a very real sense, Mountain gorillas live in large outdoor zoos and for the moment are reasonably secure there. The same may happen for select groups of Eastern gorillas and Cross River ones. And, if severely under threat of extinction, some would likely be moved to zoos and sanctuaries for protection and future reproduction if only via preserved DNA. In this way, such facilities can be seen, at least in a limited way, as fulfilling the "Ark" function often attributed to them.[57]

In the final analysis, gorillas have become our dependents. We've made them such, and left alone they inevitably would join the sixth extinction. Thanks to the efforts of the people noted above and, of course, gorillas themselves, the world at large seems to have decided in their favor, making them, to borrow from Alexander Harcourt, "lucky" compared with so many other nonhuman inhabitants of the earth, known and unknown.[58]

NOTES

1. For further details, see Andrew J. Plumptre et al., "The Current Status of Gorillas and Threats to Their Existence at the Beginning of a New Millennium," in *Gorilla Biology: A Multidisciplinary Perspective*, ed. Andrea B. Taylor and Michele L. Goldsmith (Cambridge: Cambridge University Press, 2003), 414–31.
2. Denis Ndeloh Etiendem, personal communication.
3. Richard A. Bergl and Linda Vigilant, "Genetic Analysis Reveals Population Structure and Recent Migration within the Highly Fragmented Range of the Cross River Gorilla (*Gorilla gorilla diehli*)," *Molecular Ecology* 16 (2007): 501–16.
4. John F. Oates, *Primates of West Africa: A Field Guide and Natural History* (Arlington, VA: Conservation International, 2011), 415–25.
5. John F. Oates, *Myth and Reality in the Rainforest: How Conservation Strategies Are Failing West Africa* (Berkeley: University of California Press, 1999), 129–76.
6. http://www.africanconservation.org/content/view/688/365.
7. John F. Oates et al., "The Cross River Gorilla: Natural History and Status of a Neglected and Critically Endangered Subspecies," in Taylor and Goldsmith, *Gorilla Biology*, 492.
8. Personal communication.
9. Denis Ndeloh Etiendem, personal communication.
10. For information pertaining specifically to the Congo Basin, see David S. Wilkie, "Bushmeat Trade in the Congo Basin," in *Great Apes and Humans: The Ethics of Coexistence*, ed. Benjamin B. Beck et al. (Washington, DC: Smithsonian Institution Press, 2001), 86–109. See also Dale Peterson and Karl Amman, *Eating Apes* (Berkeley: University of California Press, 2003).
11. E. Bowen-Jones and S. Pendry, "The Threat to Primates and Other Mammals from the Bushmeat Trade in Africa, and How This Threat Could Be Diminished," *Oryx* 33 (1999): 233–46.
12. Anthony L. Rose et al., *Consuming Nature: A Photo Essay on African Rainforest Exploitation* (Palos Verdes, CA: Altisima Press, 2003).
13. Peter D. Walsh et al., "Catastrophic Ape Decline in Western Equatorial Africa," *Nature* 422 (2003): 611–14.
14. Magdalena Bermejo et al., "Ebola Outbreak Killed 5000 Gorillas," *Science* 314 (2006): 1564.
15. Eric M. Leroy et al., Multiple Ebola Virus Transmission Events and Rapid Decline in Central African Wildlife," *Science* 303 (2004): 387–90.
16. Recent evidence suggests that *Falciparum* malaria, the most dangerous kind for humans, originated in gorillas and not chimpanzees, as previously thought (*New York Times*, September 28, 2010).
17. For an overview of the reserve with some good photos, see Abigail Tucker, "Primal Instinct," *Smithsonian*, November 2012, 54–63.
18. Patrick T. Mehlman, "Current Status of Wild Gorilla Populations and Their Conservation," in *Conservation in the 21st Century: Gorillas as a Case Study*, ed. Tara S. Stoinski, H. Deter Steklis, and Patrick T. Mehlman (New York: Springer, 2008), 27.

19. Ibid., 21.

20. Carla Litchfield, "Responsible Tourism: A Conservation Tool or Threat?," in Stoinski et al., *Conservation in the 21st Century*, 115.

21. Geza Teleki, "Sanctuaries for Ape Refugees," in Beck et al., *Great Apes and Humans*, 135.

22. Denis Ndeloh Etiendem, personal communication.

23. Kay H. Farmer and Amos Courage, "Sanctuaries and Reintroduction: A Role in Gorilla Conservation?," in Stoinski et al., *Conservation in the 21st Century*, 90.

24. Esteban E. Sarmiento, "Distribution, Taxonomy, Genetics, Ecology, and the Causal Links to Gorilla Survival: The Need to Develop Practical Knowledge for Gorilla Conservation," in Taylor and Goldsmith, *Gorilla Biology*, 432.

25. A detailed accounting of early menageries can be found in Gustav Loisel, *Histoire des Menageries de l'Antiquité a nos Jours*, 3 vols. (Paris: O. Doin, 1912).

26. Heini Hediger, *Man and Animal in the Zoo: Zoo Biology*, trans. Gwynne Vevers and Winwood Reade (New York: Delacorte Press, 1969), 142.

27. There's a considerable literature on the history of zoos. Readers wishing to pursue the subject will find much useful information in Wilfred Blunt, *The Ark in the Park: The Zoo in the Nineteenth Century* (London: Hamish Hamilton, 1976), and Eric Baratay and Elizabeth Hardouin-Fugier, *Zoo: A History of Zoological Gardens in the West* (London: Reaktion Books, 2002). For a look ahead, see Alexandra Zimmerman et al., *Zoos in the 21st Century: Catalysts for Conservation*? (Cambridge: Cambridge University Press, 2007).

28. David Hancocks, *A Different Nature: The Paradoxical World of Zoos and Their Uncertain Futures* (Berkeley: University of California Press, 2001), xiv.

29. A brief history of the Antwerp Zoo can be found in Ilse Segers, Marleen Huyghe, and Kris Struyf, "Zoo Antwerpen and Dierenpark Planckendael," in *Encyclopedia of the World's Zoos*, vol. 3, ed. Catherine E. Bell (Chicago: Fitzroy Dearborn Publishers, 2001), 1375–78.

30. Robert M. Yerkes, "The Mind of a Gorilla," *Genetic Psychology Monographs* 1927: 14.

31. Four videos of Gust can be found on YouTube.

32. W. Van Den Bergh, "The New Ape House at the Antwerp Zoo," unpublished manuscript, Antwerp Zoo.

33. Evelien DeGroot, "Social Integration of a Female Gorilla (*Gorilla gorilla gorilla*) into a Captive Group," unpublished manuscript, Antwerp Zoo.

34. A detailed history of conservation at the Antwerp Zoo can be found in Violette Poillard, "Les zoos et la conservation des espèces: Le cas du zoo d'Anvers," master's thesis, Université Libre de Bruxelles, 2008.

35. Angus Martin, *Gorillas in the Garden: Zoology and Zoos* (Chipping North: Surrey Beatty & Sons, 1997), 26–27.

36. Hediger, *Man and Animal in the Zoo*, 113–27.

37. For a history of zoo architecture, see Gregory T. Hyson, "Jungles of Eden: The Design of American Zoos," in *Environmentalism in Landscape Architecture*, ed. Michel Conan (Washington, DC: Dumbarton Oakes Research Library and Collection, 2000), 23–44.

38. My sources for the general history of the Philadelphia Zoo are William V. Donaldson, "An Animal Garden in Fairmont Park," unpublished manuscript, Philadelphia Zoo, 1988; Judith Ehrman and Meredith Hect, "History of the Philadelphia Zoo," unpublished manuscript, Philadelphia Zoo, 2008; Alexander Hoskins, "The Philadelphia Zoo," in Bell, *Encyclopedia of the World's Zoos*, vol. 2, 1005–11; and information from the zoo staff.

39. Kevin A. Fox, "Medical-Veterinary Team Treats Massa," *America's First Zoo* 21 (1969): 18–21.

40. A concise history of the Columbus Zoo can be found at http://www.columbuszoo.org/about_us/history. For the history as it pertains to gorillas, see Jeff Lyttle, *Gorillas in Our Midst: The Story of the Columbus Zoo Gorillas* (Columbus: Ohio State University Press, 1997), and Nancy Roe Pimm, *Colo's Story: The Life of One Grand Gorilla* (Columbus, OH: Columbus Zoological Park Association, 2011).

41. A favorable review of Aspinall exists in Nicholas Gould, "Howlett's Wild Animal Park and Port Lympe Wild Animal Park," in Bell, *Encyclopedia of the World's Zoos*, vol. 2, 589–93.

42. Unpublished manuscript, Columbus Zoo.

43. Personal communication, Columbus Zoo.

44. Ibid.

45. A brief history of Woodland Park to the turn of the century can be found in John Bierlein, "Woodland Park Zoological Gardens," in Bell, *Encyclopedia of the World's Zoos*, vol. 3, 1352–57.

46. Hancocks, *A Different Nature*, 12–30.

47. Ibid., 132.

48. Ibid., 134.

49. Personal communication, Woodland Park Zoo.

50. "A Declaration on Great Apes," in Paola Cavalieri and Peter Singer, eds., *The Great Ape Project: Equality beyond Humanity* (New York: St. Martin's Press, 1993), 4.

51. The Great Ape Debate, Project Syndicate, May 2002, http://www.utilitarian.net/singer/by/200605.

52. Michael Hutchins et al., "Rights or Welfare: A Response to the Great Ape Project, " in Beck et al., *Great Apes and Humans*, 329–66; Gary L. Francione, "The Great Ape Project: Not So Great," http://www.abolitionistapproach.com/the-great-ape-project-not-so-great.

53. George B. Schaller, *The Mountain Gorilla: Ecology and Behavior* (Chicago: University of Chicago Press, 1963), 101.

54. Good overviews of zoos and education can be found in Tara S. Stoinski et al., "Captive Apes and Zoo Education," in Beck et al., *Great Apes and Humans*, 113–32; Stephen B. Woodland, "Education History," in Bell, *Encyclopedia of the World's Zoos*, vol. 1, 397–402; and Gary V. Schwammer, "Education On-Site Programs," in Bell, *Encyclopedia of the World's Zoos*, vol. 1, 394–97.

55. *Post Standard* (Syracuse, NY), October 6, 2012.

56. Osborn's book *Our Plundered Planet* (Boston: Little, Brown and Company, 1948) was an influential stimulus to the 1960s and 1970s environmental movement. He was also a longtime board member and subsequently president of the New York

Zoological Society. For a thought-provoking current take on nature and animals unencumbered by social theory, see David Samuels, "Wild Things, Animal Nature, Human Racism and the Future of Zoos," *Harper's* 324 (2012): 33.

57. Vicki Croke, *The Modern Ark: The Story of Zoos: Past and Present* (New York: Scribner, 1997).

58. Alexander H. Harcourt, "Lucky Gorillas?," in *World Atlas of Great Apes and Their Conservation*, ed. Julian Caldecott and Lera Miles (Berkeley: University of California Press, 2005), 221.

Bibliography

Akeley, Carl E. *In Brightest Africa*. Garden City, NY: Doubleday Page & Company, 1923.

———. "Gorillas—Real and Mythical." *Natural History* 23 (1923): 428–47.

Akeley, Carl, and Mary L. Jobe. *Adventures in the African Jungle*. New York: Dodd, Mead & Company, 1950.

Akeley, Delia J. *"J. T. Jr.": The Biography of an African Monkey*. New York: The Macmillan Company, 1928.

———. *Jungle Portraits*. New York: The Macmillan Company, 1930.

Akeley, Mary L. Jobe. *Carl Akeley's Africa*. New York: Dodd, Mead & Company, 1929.

———. *The Wilderness Lives Again: Carl Akeley and the Great Adventure*. New York: Dodd, Mead & Company, 1940.

Allen, J. G. "Gorilla Hunting in Southern Nigeria." *Nigerian Field* 1 (1931): 5.

Anderson, Stephen A. *Doctor Dolittle's Delusion: Animals and the Uniqueness of Human Language*. New Haven, CT: Yale University Press, 2004.

Anonymous. "John Daniel, Civilized Gorilla." *Literary Digest*, December 10, 1922, 44–49.

———. *Traveller's Guide to the Belgian Congo and the Ruanda-Urundi*. 2nd ed. Brussels: Tourist Bureau for Belgian Congo and Ruanda-Urundi, 1956.

Aveling, Conrad, and Rosalind Aveling. "Gorilla Conservation in Zaire." *Oryx* 23 (1989): 64–70.

Ballantyne, Robert M. *The Gorilla Hunters: A Tale of the Wilds of Africa*. Philadelphia: Porter & Coates, 1861.

Baratay, Eric, and Elisabeth Hardouin-Fugier. *Zoo: A History of Zoological Gardens in the West*. London: Reaktion Books, 2002.

Barns, Thomas Alexander. *Across the Great Craterland to the Congo*. New York: Alfred A. Knopf, 1924.

———. *The Wonderland of the Eastern Congo*. London: G. P. Putnam's Sons, 1922.

Barnum, P. T. *Struggles and Triumphs: or, Forty Years' Recollections*. New York: American News Company, 1871.

Barth, Heinrich. "Analyse der Reisebeschreibung du Chaillu's *Explorations and Adventures in Equatorial Africa*, und genanere Betrachtung des in derselben enthaltenen geographischen Materials." *Zeitschrift für allgemeine Erdkunde* 10 (1861): 430–67.

Bartlett, Edward, ed. *Bartlett's Life among Wild Beasts in the "Zoo."* London: Chapman and Hall, 1900.

———. ed. *Wild Animals in Captivity*. London: Chapman and Hall, 1899.

Baumgartel, Walter. "The Gorilla Killer." *Wild Life and Sport* 2 (1961): 14–17.

———. "The Last British Gorilla." *Geographical Magazine* 32 (1959): 33–41.

———. *Up among Mountain Gorillas*. New York: Hawthorne Books, 1976.

Beck, Benjamin B., et al. *Great Apes and Humans: The Ethics of Coexistence*. Washington, DC: Smithsonian Institution Press, 2001.

Bell, Catherine E., ed. *Encyclopedia of the World's Zoos*. 3 vols. Chicago: Fitzroy Dearborn Publishers, 2001.

Bellin, Joshua D. *Framing Monsters: Fantasy Film and Social Alienation*. Carbondale: University of Southern Illinois Press, 2005.

Benchley, Belle. J. *My Friends, the Apes*. Boston: Little, Brown and Company, 1942.

———. *My Life in a Man-Made Jungle*. Boston: Little, Brown and Company, 1943.

———. "One Hundred and One Months in the Growth and Development of Mountain Gorillas." In *Wild in the City: The Best of Zoonooz*. Edited by Robert Wade. San Diego, CA: Zoological Society of San Diego, 1985, 64–67.

Bergl, Richard A., and Linda Vigilant. "Genetic Analysis Reveals Population Structure and Recent Migration within the Highly Fragmented Range of the Cross River Gorilla (*Gorilla gorilla diehli*)." *Molecular Ecology* 16 (2007): 501–16.

Bermejo, Magdalena, et al. "Ebola Outbreak Killed 5000 Gorillas." *Science* 314 (2006): 1564.

Bierlein, John. "Woodland Park Zoological Garden." In *Encyclopedia of the World's Zoos*. Vol. 3. Edited by Catherine E. Bell. Chicago: Fitzroy Dearborn Publishers, 2001, 1352–57.

Bingham, Harold C. *Gorillas in a Native Habitat*. Washington, DC: Carnegie Institute of Washington, 1932.

Blaikie, W. Garden. *The Personal Life of David Livingstone*. New York: Laymen's Missionary Movement, 1890.

Blancou, Lucien. "The Lowland Gorilla." *Animal Kingdom* 58 (1955): 162–69.

Blunt, Wilfred. *The Ark in the Park: The Zoo in the Nineteenth Century*. London: Hamish Hamilton, 1976.

Bodry-Sanders, Penelope. *Carl Akeley: Africa's Collector, Africa's Savior*. St. Paul, MN: Paragon House Publisher, 1991.

Bourne, Geoffrey Howard, and Maury Cohen. *The Gentle Giants: The Gorilla Story*. New York: Putnam and Sons, 1975.

Bowdich, T. Edward. *Mission from Cape Coast Castle to Ashantee*. London: John Murray, 1819.

Bowen-Jones, E., and S. Pendry. "The Threat to Primates and Other Mammals from the Bushmeat Trade in Africa, and How This Threat Could Be Diminished." *Oryx* 33 (1999): 233–46.

Bradley, Mary Hastings. *On the Gorilla Trail*. New York: D. Appleton and Company, 1922.

Breedon, Kristy. "Herbert Ward: Sculpture in the Circum-Atlantic World." *Visual Culture in Britain*. Special Issue, "British Culture c. 1757–1947: Global Context" 11 (2010): 265–83.

Brehm, Alfred Edmund. *Brehm's Life of Animals. "Mammalia."* Vol. 1. 3rd ed. Chicago: A. N. Marquis & Company, 1895.

Breuer, Thomas, et al. "Physical Maturation, Life-History Classes and Age-Estimates of Free-Ranging Western Gorillas—Insights from Mbeli Bai, Republic of Congo." *American Journal of Primatology* 71 (2009): 106–19.

Bucher, Henry H., Jr. "Canonization by Repetition: Paul Du Chaillu in Historiography," *Revue Française d'Histoire d'Outre-Mer* 66 (1979): 15–32.

Buisseret, David, ed. *The Oxford Companion to World Exploration*. New York: Oxford University Press, 2007.

Burbridge, Ben. *Gorilla: Tracking and Capturing the Ape-Man of Africa*. New York: The Century Co., 1928.

Burrows, Guy. *The Land of the Pigmies*. New York: Thomas Y. Crowell, 1898.

Burton, Richard F. *Two Trips to Gorilla Land and the Cataracts of the Congo*. 2 vols. London: S. Low, Marston, Low, and Searle, 1876.

Butynski, Thomas M. "Africa's Great Apes." In *Great Apes and Humans: The Ethics of Coexistence*. Edited by Benjamin B. Beck et al. Washington, DC: Smithsonian Institution Press, 2001, 3–56.

Caldecott, Julian, and Lera Miles, eds. *World Atlas of Great Apes and Their Conservation*. Berkeley: University of California Press, 2005.

Cameron, Verney Lovett. *Across Africa*. London: George Philip & Co., 1885.

Carpenter, C. R. "An Observational Study of Two Captive Mountain Gorillas." *Human Biology* 9 (1937): 175–96.

Carr, Rosamund Halsey, with Ann Howard Halsey. *Land of a Thousand Hills: My Life in Rwanda*. New York: Penguin Putnam, 2000.

Cavalieri, Paola, and Peter Singer, eds. *The Great Ape Project: Equality beyond Humanity*. New York: St. Martin's Press, 1993.

Chaillu, Paul B. Du. "Descriptions of the Habits and Distribution of the Gorilla and Other Anthropoid Apes." *Proceedings of the Boston Society of Natural History* 7 (1860): 276–77.

———. *Explorations and Adventures in Equatorial Africa*. London: J. Murray, 1861.

———. *A Journey to Ashango-Land*. London: J. Murray, 1867.

Chomsky, Noam. "Are Those Apes Really Talking?" *Time*, March 10, 1980, 50, 57.

Ciochon, Russell L., and Patricia A. Holroyd. "The Asian Origins of Anthropoidea Revisited." In *Anthropoid Origins*. Edited by John G. Fleagle and Richard F. Kay. New York: Plenum Press, 1994, 143–62.

Clark, James L. *Good Hunting, Fifty Years of Collecting and Preparing Habitat Groups for the American Museum*. Norman: University of Oklahoma Press, 1966.

Clodd, Edward. *Memories*. New York: G. P. Putnam's Sons, 1916.
Clutton-Brock, Timothy Hugh, ed. *Primate Ecology: Studies of Feeding and Ranging Behaviour in Lemurs, Monkeys, and Apes*. London: Academic Press, 1977.
Conan, Michel, ed. *Environmentalism in Landscape Architecture*. Washington, DC: Dumbarton Oakes Research Library and Collection, 2000.
Conant, Roger. "Bamboo." *Fauna* 1 (1939): 7–9.
———. "Meet the Champions." *Fauna* 3 (1941): 48.
Coolidge, Harold J., Jr. "Zoological Results of the George Vanderbilt African Expedition of 1934. Part IV,—Notes on Four Gorillas from the Sanga River Region." *Proceedings of the Academy of Natural Sciences of Philadelphia* 88 (1936): 479–501.
Corbey, Raymond. "Negotiating the Ape-Human Boundary." In *Great Apes and Humans: The Ethics of Coexistence*. Edited by Benjamin B. Beck et al. New York: Springer, 2001, 163–77.
Cousins, Don. "Gorillas—A Survey." *Oryx* 14 (1978): 374–76.
Croke, Vicki. *The Modern Ark: The Story of Zoos: Past, Present and Future*. New York: Scribner, 1997.
Cunningham, Alyse. "A Gorilla's Life in Civilization." *Bulletin of the New York Zoological Society* 24 (1921): 118–24.
Dapper, D'O. *Description de L'Afrique*. Amsterdam: Chez Wolfgang, Waesberge, Boom & van Someren, 1686.
DeGroot, Evelien. "Social Integration of a Female Gorilla (*Gorilla gorilla gorilla*) into a Captive Group." Unpublished manuscript, Antwerp Zoo, n.d.
Denis, Armand. *On Safari: The Story of My Life*. London: Collins, 1963.
Derscheid, J. M. "Notes sur les Gorillas des Volcans du Kivu (Parc National Albert)." *Extrait des Annales de la Societé Royale Zoologique de Belgique* 58 (1927): 149–59.
Donaldson, William V. "An Animal Garden in Fairmont Park." Unpublished manuscript, Philadelphia Zoo, 1988.
Donisthorpe, Jill. "Gorilla." *African Life* 1 (1957): 38–41.
———. "I Stalk Gorillas." *Personality* 30 (1958): 18–19.
———. "A Pilot Study on the Mountain Gorilla." *South African Journal of Science* 54 (1958): 195–217.
Dressman, Caroline. *A Brief History of "Susie" as Told by Her to, to Her Trainer Wm. Dressman*. Unknown publisher, 1945.
Ehrman, Judith, and Meredith Hecht. "History of the Philadelphia Zoo." Unpublished manuscript, Philadelphia Zoo, 2008.
Emlin, J. T., and George. B. Schaller. "Distribution and Status of the Mountain Gorilla 1959." *Zoologica* 45 (1960): 41–52.
———. "In the Home of the Mountain Gorilla." *Animal Kingdom* 63 (1960): 98–108.
Erb, Cynthia. *Tracking King Kong: A Hollywood Icon in World Culture*. 2nd ed. Detroit: Wayne State University Press, 2009.
Farmer, Kay H., and Amos Courage. "Sanctuaries and Reintroduction: A Role in Gorilla Conservation?" In *Conservation in the 21st Century: Gorillas as a Case Study*. Edited by Tara S. Stoinski, H. Deter Steklis, and Patrick T. Mehlman. New York: Springer, 2008, 79–106.

Fitch, W. Tecumseh, Marc D. Hauser, and Noam Chomsky. "The Evolution of Language Faculty: Clarification and Implications." *Cognition* 97 (2005): 179–210.

Fleagle, John G., and Richard F. Kay eds. *Anthropoid Origins*. New York: Plenum Press, 1994.

Ford, Henry A. "Communication." *Proceedings of the Academy of Natural Sciences of Philadelphia* 6 (1852): 30–33.

Fossey, Dian. *Gorillas in the Mist*. Boston: Houghton Mifflin, 1983.

———. "The Imperiled Mountain Gorilla." *National Geographic* 159 (1981): 501–23.

———. "Infanticide in Mountain Gorillas (*Gorilla gorilla beringei*) with Comparative Notes on Chimpanzees." In *Infanticide: Comparative and Evolutionary Perspectives*. Edited by Glenn Hausfater and Sarah Blaffer Hrdy. New York: Aldine Press, 1984, 217–35.

———. "Vocalizations of the Mountain Gorilla (*Gorilla gorilla beringei*)." *Animal Behavior* 20 (1972): 36–53.

Fossey, Dian, and Robert M. Campbell. "Making Friends with Mountain Gorillas." *National Geographic* 137 (1970): 48–67.

———. "More Years with Mountain Gorillas." *National Geographic* 140 (1971): 574–85.

Fossey, Dian, and Alexander H. Harcourt. "Feeding Ecology of Free-Ranging Mountain Gorilla (*Gorilla gorilla beringei*)." In *Primate Ecology: Studies of Feeding and Ranging Behaviour in Lemurs, Monkeys, and Apes*. Edited by Timothy Hugh Clutton-Brock. New York: Academic Press, 1977, 415–47.

Fox, Kevin A. "Medical-Veterinary Team Treats Massa." *America's First Zoo* 21 (1969): 18–21.

Fuentes, Agustín, and Linda D. Wolfe, eds. *Primates Face to Face: The Conservation Implications of Human-Nonhuman Primate Interconnections*. Cambridge: Cambridge University Press, 2002.

Garner, Harry E. *Autobiography of a Boy, from the Letters of Richard Lynch Garner*. Washington, DC: Huff Duplicating Company, 1930.

Garner, Richard L. *Apes and Monkeys: Their Life and Language*. Boston: Ginn & Company, 1900.

———. *Gorillas and Chimpanzees*. London: Osgood McIlvaine and Co., 1896.

———. "Gorillas in Their Own Jungle." *Zoological Society Bulletin* 17 (1914): 1102–4.

———. "The Simian Tongue [I]." *New Review* 4 (1891): 555–62.

———. *The Speech of Monkeys*. New York: Charles L. Webster and Company, 1892.

Gatti, Attilio. "Among the Pygmies and Gorillas." *Popular Mechanics*, September 1932, 418–21, 118A.

———. "Gorilla." *Field and Stream*," October 1932, 18–20, 66–67, 73.

———. *Tom-Toms in the Night*. London: Hutchinson & Co., 1932.

Geddis, Henry. *Gorilla*. London: Andrew Melrose, 1955.

Geoffroy-Saint-Hilaire, Isidore. "Descrition des Mammifères Nouveaux Connus de la Collection du Muséum d'Histoire Naturelle." Deuxième Supplement, *Archives de la Muséum Naturelle* 10 (1858–1861): 1–102.

Gervais, M. Paul. *Histoire naturelle des mammifères avec l'indication de leurs moeurs, et de leurs rapports avec les arts, le commerce et l'agriculture*. Paris: L. Curmer, 1854.

Goldner, Orville, and George E. Turner. *The Making of King Kong*. New York: Ballantine Books, 1975.

Goodall, Alan G. "Feeding and Ranging Behavior of a Mountain Gorilla Group (*Gorilla gorilla beringei*) in the Tshibinda-Kahuzi Region, Zaîre." In *Primate Ecology: Studies of Feeding and Ranging Behaviour in Lemurs, Monkeys, and Apes*. Edited by Thomas Hugh Clutton-Brock. London: Academic Press, 1977, 449–79.

———. *The Wandering Gorillas*. London: Collins, 1979.

Goodall, Alan G., and Colin P. Groves. "The Conservation of Eastern Gorillas." In *Primate Conservation*. Edited by His Serene Highness Prince Rainier III of Monaco and Geoffrey H. Bourne. New York: Academic Press, 1977, 599–637.

Gosse, Philip H. *The Romance of Natural History*. 1st series. London: John Nisbet and Company, 1861.

Gottesman, Ronald, and Henry Geduld, eds. *The Girl in the Hairy Paw: King Kong as Myth, Movie, and Monster*. New York: Avon Books, 1976.

Gould, Nicholas. Howletts Wild Animal Park and Port Limpe Wild Animal Park." In *Encyclopedia of the World's Zoo*. Vol. 2. Edited by Catherine E. Bell. Chicago: Fitzroy Dearborn Publishers, 2001, 589–93.

Graham, Charles E., ed. *Reproductive Biology of the Great Apes*. New York: Academic Press, 1981.

Greeley, Adolphus W. *Men of Achievement, Explorers and Travelers*. New York: Charles Scribner's Sons, 1893.

Gregory, W. K., and H. C. Raven. *In Quest of Gorillas*. New Bedford, MA: Darwin Press, 1937.

Grogan, Ewart S., and Arthur H. Sharp. *From the Cape to Cairo*. 2nd ed. London: Thomas Nelson & Sons, 1920.

Groves, C. P. "Distribution and Place of Origin of the Gorilla." *Man*, n.s. 6 (1971); 44–51.

———. "A History of Gorilla Taxonomy." In *Gorilla Biology: A Multidisciplinary Perspective*. Edited by Andrea B. Taylor and Michele L. Goldsmith. Cambridge: Cambridge University Press, 2003, 15–34.

———. "A Note on the Affinities of the Ebo Forest Gorilla." *Gorilla Journal* 31 (2005): 19–21.

Hancocks, David. *A Different Nature: The Paradoxical World of Zoos and Their Uncertain Futures*. Berkeley: University of California Press, 2001.

Haraway, Donna J. *When Species Meet*. Minneapolis: University of Minnesota Press, 2008.

Harcourt, Alexander H. "An Introductory Perspective: Gorilla Conservation." In *Gorilla Biology: A Multidisciplinary Perspective*. Edited by Andrea B. Taylor and Michele L. Goldsmith. Cambridge: Cambridge University Press, 2003, 407–13.

———. "Lucky Gorillas?" In *World Atlas of Great Apes and Their Conservation*. Edited by Julian Caldecott and Lera Miles. Berkeley: University of California Press, 2005, 221.

Harcourt, Alexander H., and Dian Fossey, "The Virunga Gorilla: Decline of an Island Population." *African Journal of Ecology* 19 (1981): 83–97.

Harcourt, Alexander H., and Kelly J. Stewart. *Gorilla Society: Conflict, Compromise and Cooperation between the Sexes.* Chicago: University of Chicago Press, 2007.

Harcourt, Alexander H., Kelly Stewart, and Dian Fossey. "Gorilla Reproduction in the Wild." In *Reproductive Biology of the Great Apes*. Edited by Charles E. Graham. New York: Academic Press, 1981, 265–79.

———. "Male Emigration and Female Transfer in Wild Mountain Gorilla." *Nature* 263 (1976): 226–27.

Hauser, Marc D., Noam Chomsky, and W. Tecumseh Fitch. "The Faculty of Language: What Is It, Who Has It, and How Did It Evolve?" *Science* 298 (2002): 1569–79.

Hausfater, Glenn, and Sarah Blaffer Hrdy, eds. *Infanticide: Comparative and Evolutionary Perspectives*. New York: Aldine, 1984.

Haver, Ronald. *David O. Selznick's Hollywood.* New York: Alfred A. Knopf, 1980.

Hayes, Harold T. P. *The Dark Romance of Dian Fossey*. New York: Simon and Schuster, 1990.

Hediger, Heini. *Man and Animal in the Zoo: Zoo Biology*. Translated by Gwynne Vevers and Winwood Reade. New York: Delacorte Press, 1969.

Henderson, J. Y. *Circus Doctor*. Boston: Little, Brown and Company, 1973.

His Serene Highness Prince Rainier III of Monaco and Geoffrey H. Bourne, eds. *Primate Conservation.* New York: Academic Press, 1977.

Hoskins, Alexander. "The Philadelphia Zoo." In *Encyclopedia of the World's Zoos*. Vol. 2. Edited by Catherine E. Bell. Chicago: Fitzroy Dearborn Publishers, 2001, 1005–11.

Hoyt A. M. *Toto and I: A Gorilla in the Family*. Philadelphia: Lippincott, 1941.

Hutchins, Michael, et al. "Rights or Welfare: A Response to the Great Ape Project." In *Great Apes and Humans: The Ethics of Coexistence*. Edited by Benjamin B. Beck et al. Washington, DC: Smithsonian Institution Press, 2001, 329–66.

Huxley, Thomas H. *Man's Place in Nature and Other Anthropological Essays*. New York: D. Appleton and Company, 1909.

Hyson, Gregory T. "Jungles of Eden: The Design of American Zoos." In *Environmentalism in Landscape Architecture*. Edited by Michel Conan. Washington, DC: Dumbarton Oaks Research Library and Collection, 2000, 23–44.

Imanishi, Kinji. "Gorilla: A Preliminary Survey in 1958." *Primates* 1 (1958): 73–78.

Imperato, Pascal J., and Eleanor M. Imperato. *They Married Adventure: The Wandering Lives of Martin and Osa Johnson*. New Brunswick, NJ: Rutgers University Press, 1992.

Jackendoff, Ray, and Steven Pinker. "The Nature of the Language Faculty and Its Implications for Evolution of Language (reply to Fitch, Hauser, and Chomsky)." *Cognition* 97 (2005): 211–25.

Janson, Horst W. *Apes and Ape Lore in the Middle Ages and the Renaissance*. London: Warburg Institute and University of London, 1952.

Jenkins, Martin. "Evolution, Dispersal, and Discovery of the Great Apes." In *World Atlas of Great Apes and Their Conservation*. Edited by Julian Caldecott and Lera Miles. Berkeley: University of California Press, 2005, 13–28.

Jenks, Albert E. "Bulu Knowledge of the Gorilla." *American Anthropologist* 13 (1911): 56–64.

Johnson, Martin E. *Camera Trails in Africa*. New York: Grosset & Dunlap, 1924.

———. *Congorilla*. New York: Brewer, Warren & Putnam, 1931.

Johnston, Sir Harry. *George Grenfell and the Congo*. 2 vols. London: Hutchinson & Co., 1908.

Jones, Jeanette Eileen. "'Gorilla Trails in Paradise': Carl Akeley, Mary Bradley, and the American Search for the Missing Link." *Journal of American Culture* 29 (2006): 321–36.

Kawai, Masao, and Hiraki Mizuhara, "An Ecological Study of the Wild Mountain Gorilla (*Gorilla gorilla beringei*)." *Primates* 2 (1959): 1–42.

Keith, Sir Arthur. "An Introduction to the Study of Anthropoid Apes. 1. The Gorilla." *Natural Science* 9 (1896): 26–37.

Kirk, Jay. *Kingdom under Glass: A Tale of Obsession, Adventure, and One Man's Quest to Preserve the World's Great Animals*. New York: Henry Holt and Company, 2010.

Klieman, Kairn A. *"The Pygmies Were Our Compass": Bantu and Batwa in the History of West Central Africa, Early Times to 1900 C.E.* Portsmouth, NH: Heinemann, 2003.

Kuklick, Henrika, and Robert Kohler, eds. "Science in the Field." *Osiris*, 2nd ser., 11 (1996).

Lang, Ernst M. *Goma, the Baby Gorilla*. Translated by Edmund Fisher. London: Victor Gollancz, 1962.

Leroy, Eric M., et al. "Multiple Ebola Virus Transmission Events and Rapid Decline in Central African Wildlife." *Science* 303 (2004): 387–90.

Lindsey, Jennifer. *The Great Apes*. New York: Metro Books, 1999.

Lintz, Gertrude Davies. *Animals Are My Hobby*. New York: Robert M. McBride & Company, 1942.

Litchfield, Carla A. "Responsible Tourism: A Conservation Tool or Threat?" In *Conservation in the 21st Century: Gorillas as a Case Study*. Edited by Tara S. Stoinski, H. D. Deter Steklis, and Patrick T. Mehlman. New York: Springer, 2008, 107–35.

Loisel, Gustav. *Histoire des Menageries de l'Antiquité a nos Jours*. 3 vols. Paris: O. Doin, 1912.

Lynn, Barbara. *The Heyday of Natural History, 1820–1870*. Garden City, NY: Doubleday & Company, 1980.

Lyttle, Jeff. *Gorillas in Our Midst: The Story of the Columbus Zoo Gorillas*. Columbus: Ohio State University Press, 1997.

Mandelstam Joel. "Du Chaillu's Stuffed Gorillas and the Savants of the British Museum." *Notes and Records of the Royal Society of London* 48 (1994): 227–45.

Martin, Angus. *Gorillas in the Garden: Zoology and Zoos*. Chipping Norton: Surrey Beatty & Sons, 1997.

Matschie, Paul. "Uber einen Gorilla aus Deutsch-Ostafrika." *Sitzungsberichte der Gesellschaft Naturforschender Freunde Berlin* 1903: 253–59.

Mayne, Judith. "King Kong and the Ideology of the Spectacle." *Quarterly Review of Film Studies* 1 (1976): 384–90.

McCook, Stuart. "It May Be Truth, but It Is Not Evidence: Paul Du Chaillu and Legitimation of Evidence in the Field Sciences." In "Science in the Field." Edited by Henrika Kuklick and Robert Kohler. *Osiris*, 2nd ser., 11 (1996): 177–97.

McFarland, K. L. "Ecology of Cross River Gorillas (*Gorilla gorilla diehli*) on Afi Mountain, Cross River State, Nigeria." PhD diss., City University of New York, 2007.

Mehlman, Patrick T. "Current Status of Wild Gorilla Populations and Their Conservation." In *Conservation in the 21st Century: Gorillas as a Case Study*. Edited by Tara S. Stoinski, H. Deter Steklis, and Patrick T. Mehlman. New York: Springer, 2008, 3–54.

Merfield, Fred G., and Harry Miller. *Gorilla Hunter*. New York: Farrar, Straus and Cudahy, 1956.

———. *Gorillas Were My Neighbors*. London: Longman, Green, 1956.

Milton, Oliver. "The Last Stronghold of the Mountain Gorilla in East Africa." *Animal Kingdom* 60 (1957): 58–61.

Morris, Ramona, and Desmond Morris. *Men and Apes*. New York: McGraw-Hill Book Co., 1966.

Mowat, Farley. *Woman in the Mists: The Story of Dian Fossey and the Mountain Gorillas of Africa*. New York: Warner Books, 1987.

Nassau, Robert Hamill. *In an Elephant Corral*. New York: Neal Publishing Company, 1912.

Newman, James L. "Discovering Gorillas: The Journey from Mythic to Real." *Terrae Incognitae* 38 (2006): 36–54.

Newman, Ken. "Saza Chief." *African Wildlife* 13 (1959): 137–42.

Nichols, Michael, and George B. Schaller. *Gorilla Struggle for Survival in the Virungas*. New York: Aperture Foundation, 1989.

Nienaber, Georgianne. *Gorilla Dreams: The Legacy of Dian Fossey*. New York: iUniverse Inc., 2006.

Noell, Anna Mae. *Gorilla Show*. Tarpon Springs, FL: Noell's Ark Publisher, 1979.

North, Henry Ringling, and Alden Hatch. *The Circus Kings: Our Ringling Family Story*. Garden City, NY: Doubleday & Company, 1960.

Nott, John Fortune. *Wild Animals Photographed and Described*. London: Sampson Low, Marston, Searle, & Rivington, 1886.

Oates, John F. *Myth and Reality in the Rain Forest: How Conservation Strategies Are Failing in West Africa*. Berkeley: University of California Press, 1999.

———. *Primates of West Africa: A Field Guide and Natural History*. Arlington, VA: Conservation International, 2011.

Oates, John F., et al. "The Cross River Gorilla: Natural History and Status of a Neglected and Critically Endangered Subspecies." In *Gorilla Biology: A Multidisciplinary Perspective*. Edited by Andrea B. Taylor and Michele L. Goldsmith. Cambridge: Cambridge University Press, 2003, 472–97.

Oikonomides, A. N., and M. C. J. Miller. *Hanno the Carthaginian: Periplus, or* Circumnavigation of Africa. 3rd ed. Chicago: ARES Publishers, 1995.

Olds, Elizabeth Fagg. *Women of the Four Winds*. Boston: Houghton Mifflin Company, 1985.

Onslow, John. *Captured by a Gorilla, etc*. Philadelphia: C. W. Alexander, 1867.

Osborn, Henry Fairfield, Jr. *Our Plundered Planet*. Boston: Little, Brown and Company, 1948.

Osborn, Rosalie. "Observations on the Behavior of the Mountain Gorilla." *Primates* 10 (1963): 29–37.

Owen, Richard. "On the Gorilla (*Troglodytes gorilla* Sav.)." *Proceedings of the Zoological Society of London* (1859): 1–23.

Parnell, Richard. "Forest Clearings: A Window into the World of Gorillas." In *World Atlas of Great Apes and Their Conservation*. Edited by Julian Caldecott and Lera Miles. Berkeley: University of California Press, 2005, 113–14.

Patterson, Francine. *Koko-Love! Conversations with a Signing Gorilla*. New York: Dutton Children's Books, 1999.

———. *Koko's Kitten*. New York: Scholastic, 1985.

———. *Koko's Story*. New York: Scholastic, 1987.

Patterson, Francine, and Wendy Gordon. "The Case for the Personhood of Gorillas." In *The Great Ape Project*. Edited by Paola Cavalieri and Peter Singer. New York: St. Martin's Press, 1993, 58–77.

Patterson, Francine, and Eugene Linden. *The Education of Koko*. New York: Holt, Rinehart and Winston, 1981.

Patterson, K. David. "Paul B. Du Chaillu and the Exploration of Gabon 1855–1865." *International Journal of African Historical Studies* 7 (1974): 647–67.

Peterson, Dale, and Karl Amman. *Eating Apes*. Berkeley: University of California Press, 2003.

Pimm, Nancy Roe. *Colo's Story: The Life of One Grand Gorilla*. Columbus, OH: Columbus Zoological Park Association, 2011.

Pinker, Steven. *The Language Instinct*. New York: William Morrow and Company, 1994.

Pitman, Charles R. S. *A Game Warden among the Charges*. London: Nisbet & Co., 1931.

Plowden, Gene. *Gargantua: Circus Star of the Century*. Miami, FL: E. A. Seeman Publishing, 1972.

Plumptre, Andrew J., and Elizabeth A. Williamson, "Conservation-Oriented Research in the Virunga Region." In *Mountain Gorillas: Three Decades of Research at Karisoke*. Edited by Martha M. Robbins, Pascale Sicotte, and Kelly J. Stewart. Cambridge: Cambridge University Press, 2001, 361–89.

Plumptre, Andrew J., et al. "The Current Status of Gorillas and Threats to Their Existence at the Beginning of a New Millennium." In *Gorilla Biology: A Multidisciplinary Perspective*. Edited by Andrea B. Taylor and Michele L. Goldsmith. Cambridge: Cambridge University Press, 2003, 414–31.

Pouillard, Violette. "Les zoos et la conservation des espèces: Le cas du zoo d'Anvers." Master's thesis, Université Libre de Bruxelles, 2008.

Poynter, Margaret. *The Zoo Lady: Belle Benchley and the San Diego Zoo*. Minneapolis: Dillon Press, 1980.

Premack, David, and Ann James Premack. *The Mind of an Ape*. New York: W. W. Norton & Company, 1983.

Pretorius, P. J. *Jungle Man: Autobiography of Major P. J. Pretorius*. New York: E. P. Dutton & Company, 1948.

Prince William of Sweden. *Among Pygmies and Gorillas*. New York: E. P. Dutton and Company, n.d.

Radick, Gregory. *The Simian Tongue: The Long Debate about Animal Language*. Chicago: University of Chicago Press, 2007.

Raven, H. C. "Gorilla: The Greatest of All Apes." *Natural History* 31 (1931): 231–42.

Ravenstein, E. G., ed. *The Strange Adventures of Andrew Battell of Leigh in Angola and the Adjoining Regions*. Reprinted from "Purchase His Pilgrimes." London: The Hakluyt Society, 2nd series, vol. 6, 1901.

Reade, W. Winwood. "The Habits of the Gorilla." *American Naturalist* 1 (1867): 177–80.

———. *Savage Africa*. New York: Harper & Brothers, Publisher, 1864.

Rhen, A. G. "Zoological Results of the George Vanderbilt African Expedition of 1934. Part I,—Introduction and Itinerary." *Proceedings of the Academy of Natural Sciences of Philadelphia* 88 (1936): 1–14.

Riopelle, Arthur J. "Growing Up with Snowflake." *National Geographic* 138 (1970): 491–502.

———. "'Snowflake': The World's First White Gorilla." *National Geographic* 131 (1967): 44–48.

Robbins, Martha M., Pascale Sicotte, and Kelly J. Stewart, eds. *Mountain Gorillas: Three Decades of Research at Karisoke*. Cambridge: Cambridge University Press, 2001.

Ron, T. "The Majombe Forest in Cabinda: Conservation Efforts, 2000–2004." *Gorilla Journal* 30 (2005): 18–21.

Rose, Anthony L., et al. *Consuming Nature: A Photo Essay on African Rainforest Exploitation*. Palos Verdes, CA: Altisima Press, 2003.

Rosenthal, Mark, Carole Tauber, and Edward Uhlir. *The Ark in the Park: The Story of Lincoln Park Zoo*. Urbana: University of Illinois Press, 2003.

Russell, Mrs. Charles E. B. *My Monkey Friends*. 2nd ed. London: Adam & Charles Black, 1948.

Samuels, David. "Wild Things, Animal Nature, Human Racism, and the Future of Zoos." *Harper's Magazine* 324 (2012): 28–42.

Sarmiento, Esteban E. "Distribution, Taxonomy, Genetics, Ecology, and the Causal Links to Gorilla Survival: The Need to Develop Practical Knowledge for Gorilla Conservation." In *Gorilla Biology: A Multidisciplinary Perspective*. Edited by Andrea B. Taylor and Michele L. Goldsmith. Cambridge: Cambridge University Press, 2003, 432–71.

Sarmiento Esteban E., and John F. Oates. "The Cross River Gorillas: A Distinct Subspecies, *Gorilla gorilla diehli Matchie* 1904." *American Museum Novitiates* 3304 (2000): 2–55.

Savage, Thomas S., and Jeffries Wyman. "Notice of the External Characters and Habits of *Troglodytes gorilla*, a New Species of Orang from the Gaboon River; Osteology of the Same." *Boston Journal of Natural History* 5 (1847): 417–41.

Savage-Rumbaugh, Sue, Stuart G. Shanker, and Talbot J. Taylor. *Apes, Language, and the Human Mind*. New York: Oxford University Press, 1998.

Schaller, George B. "Gentle Gorillas, Turbulent Times." *National Geographic* 188 (1995): 58–69.

———. *The Mountain Gorilla: Ecology and Behavior*. Chicago: University of Chicago Press, 1963.

———. *The Year of the Gorilla*. Chicago: University of Chicago Press, 1964.

Scharf, Thomas L. "Benchley, Belle Jennings 1882–1973 American Zoo Director." In *Encyclopedia of the World's Zoos*. Vol. 1. Edited by Catherine E. Bell. Chicago: Fitzroy Dearborn Publishers, 2001, 121–23.

Scherren, Henry. *The Zoological Society of London: A Sketch of Its Foundation and Development*. London: Cassell and Company Limited, 1905.

Schleier, Merrill. "The Empire State Building, Working Class Masculinity, and *King Kong*." *Mosaic* 41 (2008): 29–54.

Schwammer, Gary V. "Education On-Site Programs." In *Encyclopedia of the World's Zoos*. Vol. 1. Edited by Catherine E. Bell. Chicago: Fitzroy Dearborn Publishers, 2001, 397–402.

Segers, Ilse, Marleen Huyghe, and Kris Struyf. "Zoo Antwerpen and Dierenpark Planckendael." In *Encyclopedia of the World's Zoos*. Vol. 3. Edited by Catherine E. Bell. Chicago: Fitzroy Dearborn Publishers, 2001, 1375–78.

Shapiro, Kenneth, and Margo DeMello. "The State of Human-Animal Studies." *Society and Animals* 18 (2010): 1–17.

Sharp, N. A. "Notes on the Gorilla." *Proceedings of the Zoological Society of London* 97 (1927): 1006–9.

Sicotte, Pascal, and Prosper Uwengeli. "Reflections on the Concept of Nature and Gorillas in Rwanda: Implications for Conservation." In *Primates Face to Face: The Conservation Implications of Human-Nonhuman Primate Interconnections*. Edited by Augustín Fuentes and Linda D. Wolfe. Cambridge: Cambridge University Press, 2002, 163–81.

Singer, Peter. *Animal Liberation*. New York: Random House, 1975.

Siroto, Leon, and Kathleen Berrin. *East of the Atlantic West of the Congo: Art from Equatorial Africa*. Seattle: University of Washington Press, 1995.

Sparks, Richard D. "Congo: A Personality." *Field and Stream*, January 1926, 18–20, 72–73.

Speke, John Hanning. *Journal of the Discovery of the Nile*. New York: Harper & Brothers Publishers, 1864.

Stanley, Henry M. *My Dark Companions and Their Strange Stories*. New York: Charles Scribner's Sons, 1893.

Stoinski, Tara S., H. Deter Steklis, and Patrick T. Mehlman, eds. *Conservation in the 21st Century: Gorillas as a Case Study*. New York: Springer, 2008.

Stoinski, Tara S., et al. "Captive Apes and Zoo Education." In *Great Apes and Humans: The Ethics of Coexistence*. Edited by Benjamin B. Beck et al. Washington, DC: Smithsonian Institution Press, 2001, 113–32.

Stokes, Emma J., Richard Parnell, and Claudia Olejniczak. "Female Dispersal and Reproductive Success in Wild Western Lowland Gorillas (*Gorilla gorilla gorilla*)." *Behavioral Ecology and Sociobiology* 54 (2003): 329–39.

Taylor, Andrea B., and Michele L. Goldsmith, eds. *Gorilla Biology: A Multidisciplinary Perspective*. Cambridge: Cambridge University Press, 2003.

Teleki, Geza. "Sanctuaries for Ape Refugees." In *Great Apes and Humans: The Ethics of Coexistence.* Edited by Benjamin B. Beck et al. Washington, DC: Smithsonian Institution Press, 2001, 133–49.

Terrace, Herbert S. *Nim*. New York: Alfred A. Knopf, 1979.

Terrace, Herbert S., et al. "Can an Ape Create a Sentence?" *Science* 206 (1979): 891–902.

Tyler, Tom, and Manuela Rossini, eds. *Animal Encounters*. Leiden: Brill, 2009.

Wade, Robert, ed. *Wild in the City: The Best of Zoonooz*. San Diego, CA: Zoological Society of San Diego, 1985.

Van Den Bergh, W. "The New Ape House at the Antwerp Zoo." Unpublished manuscript, Antwerp Zoo, n.d.

Vaucaire, Michel. *Paul Du Chaillu: Gorilla Hunter*. New York: Harper & Brothers, 1930.

Walker, R. B. "M. Du Chaillu and His Book." *Athenaeum* 21 (September 1861): 374.

Walsh, Peter D., et al. "Catastrophic Ape Decline in Western Equatorial Africa." *Nature* 422 (2003): 611–14.

Waterton, Charles. *Essays on Natural History*. London: Longman, Brown, Green, Longmans, and Roberts, 1857.

Weber, Bill, and Amy Vedder. *In the Kingdom of Gorillas: Fragile Species in a Dangerous Land*. New York: Simon & Schuster, 2001.

Wilkie, David S. "Bushmeat Trade in the Congo Basin." In *Great Apes and Humans: The Ethics of Coexistence.* Edited by Benjamin B. Beck et al. Washington, DC: Smithsonian Institution Press, 2001, 86–109.

Willoughby, D. B. *All about Gorillas*. Cranbury, NJ: A. S. Barnes, 1979.

Woodland, Stephen B. "Education History." In *Encyclopedia of the World's Zoos*. Vol. 1. Edited by Catherine E. Bell. Chicago: Fitzroy Dearborn Publishers, 2001, 394–97.

Woods, Vanessa. *Bonobo Handshake: A Memoir of Love and Adventure in the Congo*. New York: Gotham Books, 2010.

Wordsworth (Donisthorpe), Jill. *Gorilla Mountain*. London: Lutterworth Press, 1961.

Yamagiwa, Juishi, and John Kahekwa. "Dispersal Patterns, Group Structure, and Reproductive Parameters of Eastern Lowland Gorillas at Kahuzi in the Absence of Infanticide." In *Mountain Gorillas: Three Decades of Research at Karisoke*. Edited by Martha M. Robbins, Pascale Sicotte, and Kelly J. Stewart. Cambridge: Cambridge University Press, 2001, 89–122.

Yamagiwa, Juishi, and N. Mwanza. "Day-Journey Length and Daily Diet of Solitary Male Gorillas in Lowland and Highland Habitats." *International Journal of Primatology* 15 (1994): 207–24.

Yamagiwa, Juishi, et al. "Seasonal Change in the Composition of the Diet of Eastern Lowland Gorillas." *Primates* 35 (1994): 1–14.

Yerkes, Robert M. *Almost Human*. New York: The Century Co., 1925.

———. "The Mind of a Gorilla." *Genetic Psychology Monographs* 1927: 1–197; Part II, 1927: 381–582; Part III in *Comparative Psychology Monographs* 5 (1929): 1–92.

Yerkes, Robert M., and Ada W. Yerkes. *The Great Apes: A Study of Anthropoid Life*. New Haven, CT: Yale University Press, 1929.

Zahl, Paul A. "Face to Face with Gorillas in Central Africa." *National Geographic* 117 (1960): 114–37.

Zimmermann, Alexandra, et al. *Zoos in the 21st Century: Catalysts for Conservation?* Cambridge: Cambridge University Press, 2007.

Zoological Society of Philadelphia. "A Garden in Fairmont Park." Unpublished manuscript, Philadelphia Zoo, 1988.

Index

About the Author

James L. Newman is emeritus professor of geography in the Maxwell School at Syracuse University. He's an Africa specialist whose previous books include *The Peopling of Africa: A Geographic Interpretation*, *Imperial Footprints: Henry Morton Stanley's African Journeys*, and *Paths without Glory: Richard Francis Burton in Africa*.